IFCoLog Journal of Logics and their Applications

Volume 3, Number 2

August 2016

Disclaimer

Statements of fact and opinion in the articles in IfCoLog Journal of Logics and their Applications are those of the respective authors and contributors and not of the IfCoLog Journal of Logics and their Applications or of College Publications. Neither College Publications nor the IfCoLog Journal of Logics and their Applications make any representation, express or implied, in respect of the accuracy of the material in this journal and cannot accept any legal responsibility or liability for any errors or omissions that may be made. The reader should make his/her own evaluation as to the appropriateness or otherwise of any experimental technique described.

© Individual authors and College Publications 2016
All rights reserved.

ISBN 978-1-84890-221-3
ISSN (E) 2055-3714
ISSN (P) 2055 3706

College Publications
Scientific Director: Dov Gabbay
Managing Director: Jane Spurr

http://www.collegepublications.co.uk

Printed by Lightning Source, Milton Keynes, UK

All rights reserved. No part of this publication may be reproduced, stored in a retrieval system or transmitted in any form, or by any means, electronic, mechanical, photocopying, recording or otherwise without prior permission, in writing, from the publisher.

Editorial Board

Editors-in-Chief
Dov M. Gabbay and Jörg Siekmann

Marcello D'Agostino	Melvin Fitting	Henri Prade
Natasha Alechina	Michael Gabbay	David Pym
Sandra Alves	Murdoch Gabbay	Ruy de Queiroz
Arnon Avron	Thomas F. Gordon	Ram Ramanujam
Jan Broersen	Wesley H. Holliday	Chrtian Retoré
Martin Caminada	Sara Kalvala	Ulrike Sattler
Balder ten Cate	Shalom Lappin	Jörg Siekmann
Agata Ciabttoni	Beishui Liao	Jane Spurr
Robin Cooper	David Makinson	Kaile Su
Luis Farinas del Cerro	George Metcalfe	Leon van der Torre
Esther David	Claudia Nalon	Yde Venema
Didier Dubois	Valeria de Paiva	Rineke Verbrugge
PM Dung	Jeff Paris	Heinrich Wansing
Amy Felty	David Pearce	Jef Wijsen
David Fernandez Duque	Brigitte Pientka	John Woods
Jan van Eijck	Elaine Pimentel	Michael Wooldridge

Scope and Submissions

This journal considers submission in all areas of pure and applied logic, including:

pure logical systems
proof theory
constructive logic
categorical logic
modal and temporal logic
model theory
recursion theory
type theory
nominal theory
nonclassical logics
nonmonotonic logic
numerical and uncertainty reasoning
logic and AI
foundations of logic programming
belief revision
systems of knowledge and belief
logics and semantics of programming
specification and verification
agent theory
databases

dynamic logic
quantum logic
algebraic logic
logic and cognition
probabilistic logic
logic and networks
neuro-logical systems
complexity
argumentation theory
logic and computation
logic and language
logic engineering
knowledge-based systems
automated reasoning
knowledge representation
logic in hardware and VLSI
natural language
concurrent computation
planning

This journal will also consider papers on the application of logic in other subject areas: philosophy, cognitive science, physics etc. provided they have some formal content.

Submissions should be sent to Jane Spurr (jane.spurr@kcl.ac.uk) as a pdf file, preferably compiled in LaTeX using the IFCoLog class file.

CONTENTS

Editorial Preface 175
   Matthias Thimm and Dov Gabbay

ARTICLES

**Computing or Estimating Extensions' Probabilities over Structured
   Probabilistic Argumentation Frameworks** 177
   Fazzinga et al.

**Against Narrow Optimization and Short Horizons:
   An Argument-based, Path Planning, and Variable Multiattribute Model
   for Decision and Risk** 201
   Ronald P. Loui

Introducing Bayesian Argumentation Networks 241
   Dov Gabbay and Odinaldo Rodrigues

EDITORIAL PREFACE

MATTHIAS THIMM
Universität Koblenz-Landau, Germany
thimm@uni-koblenz.de

DOV GABBAY
*Department of Informatics, King's College London,
Ashkelon Academic College, Israel,
Bar Ilan University, Ramat Gan, Israel
University of Luxembourg, Luxembourg.*
dov.gabbay@kcl.ac.uk

Approaches to computational argumentation have gained much attention within Artificial Intelligence. They deal with the interaction of arguments through attacks and how acceptable conclusions can be drawn from conflicting sets of arguments. These approaches provide an intuitive representation of many other non-monotonic and commonsense reasoning techniques. Research in computational argumentation is either based on the abstract approach, which focuses on the interaction of arguments and treats arguments as atomic entities, or on the structured approach, where arguments are composed of formulas of an underlying logic and the attack relation derives from conflict properties of this logic.

Recently, the augmentation of computational models of argument with quantitative forms of uncertainty has become an active endeavor within the community. While classical computational argumentation approaches (both abstract and structured) provide a qualitative form of defeasible reasoning, adding quantitative uncertainty (either in the form of probabilities, fuzzy values, or other weights) increases the expressiveness of the formalisms and provides a more natural way of dealing with uncertainty.

This special issue "Probabilistic and other Quantitative Approaches to Computational Argumentation" surveys the current state of the art on integrating quantitative uncertainty in computational models of argumentation. It features three works addressing different aspects from this new field at the intersection of qualitative and quantitative reasoning

The paper "Computing or Estimating Extension's Probabilities over Structured Probabilistic Argumentation Frameworks" by Bettina Fazzinga, Sergio Flesca,

Francesco Parisi, and Adriana Pietramala deals with abstract argumentation frameworks extended by a structured account to augment this framework with probabilities. Algorithms for computing and approximating probabilities in this setting are devised and empirically evaluated. A central result is that, despite its higher computational complexity, the exact algorithm outperforms the approximate one in certain cases.

In "Against Narrow Optimization and Short Horizons: An Argument-based, Path Planning, and Multiattribute Model for Decision and Risk" Ronald P. Loui introduces a conceptual model for representing arguments, augmented by probabilistic information, for the analysis of decision and risk. Through several examples it is shown that the use of this novel framework allows a deeper and more appropriate representation of complex decision problems involving risk assessment. The paper provides an in-depth discussion of the issues inherent to those scenarios and brings up some thought-provoking directions for future works in this area.

Finally, the work "Introducing Bayesian Argumentation Networks" by Dov M. Gabbay and Odinaldo Rodriguez presents a representation of Bayesian Networks in argumentation frameworks extended with numerical values. This translation allows for a formal comparison between these two formalisms and, besides others, provides an interpretation of cyclic Bayesian Networks through argumentative terms.

COMPUTING OR ESTIMATING EXTENSION'S PROBABILITIES OVER STRUCTURED PROBABILISTIC ARGUMENTATION FRAMEWORKS

BETTINA FAZZINGA
ICAR-CNR, Rende (CS), Italy
`fazzinga@icar.cnr.it`

SERGIO FLESCA
DIMES, University of Calabria, Rende (CS), Italy
`flesca@dimes.unical.it`

FRANCESCO PARISI
DIMES, University of Calabria, Rende (CS), Italy
`fparisi@dimes.unical.it`

ADRIANA PIETRAMALA
DISCo, University of Milano-Bicocca, Milano, Italy
`adriana.pietramala@disco.unimib.it`

Abstract

Probabilistic argumentation combines Dung's abstract argumentation framework with probability theory in order to model uncertainty in argumentation. In this setting, we address the fundamental problem of computing the probability that a set of arguments is an *extension* according to a given semantics over structured probabilistic argumentation frameworks. We focus on the most popular semantics (i.e., *admissible*, *stable*, *complete*, *grounded*, and *preferred*), for which the problem of computing extension's probabilities over structured probabilistic argumentation frameworks was shown to be $FP^{\#P}$-complete. Our aim is that of experimentally establishing when, due to the complexity of the problem and the size of the structured probabilistic argumentation framework, estimating the extension's probabilities is preferable to computing it (as computing the probability cannot be done in reasonable time). To do this, we devise two algorithms: the naive one, which computes the extension's probabilities, and the Monte-Carlo simulation one, which estimates the extension's probabilities, and evaluate both algorithms over two datasets to compare their efficiency.

Keywords: Probabilistic Argumentation Framework, Abstract Argumentation, Extension, Semantics, Monte-Carlo Simulation

1 Introduction

Argumentation allows disputes to be modeled, which arise between two or more parties, each of them providing arguments to assert their reasons. Although argumentation is strongly related to philosophy and law, it has gained remarkable interest in AI as a reasoning model for representing dialogues, making decisions, and handling inconsistency and uncertainty [9, 10, 39]. In this context, the abstract argumentation framework (AAF) [13] has been proposed, which is a powerful but simple way for modeling disputes. An AAF is a pair $\langle A, D \rangle$ consisting of a set A of *arguments*, and of a binary relation D over A, called *defeat* (or, equivalently, *attack*) relation. Basically, an argument is an abstract entity that may attack and/or be attacked by other arguments. Several semantics for AAFs, such as *admissible*, *stable*, *preferred*, and others, have been proposed [13, 14, 7] to identify "reasonable" sets of arguments, called *extensions*. Basically, each of these semantics corresponds to some properties which "certify" whether a set of arguments can be profitably used to support a point of view in a discussion. For instance, a set S of arguments is an extension according to the admissible semantics if it has two properties: it is conflict-free (that is, there is no defeat between arguments in S), and every argument (outside S) attacking an argument in S is counterattacked by an argument in S. Intuitively enough, the fact that a set is an extension according to the admissible semantics means that, using the arguments in S, you do not contradict yourself, and you can rebut to anyone who uses any of the arguments outside S to contradict yours. The complexity of the problem of verifying whether a given set of arguments is an extension according to a semantics was addressed in [19, 17, 20].

Example 1. *Consider the following scenario (inspired by an example in [27]), where the arguments are:*

  a: Prescribe John diuretics since John has hypertension,

  b: Prescribe John beta blockers since John has hypertension,

  c: John has emphysema.

Herein, since taking both diuretics and beta blockers at the same time is generally not recommended, argument a may attack argument b, and vice versa. Moreover, argument c may attack b since it is generally not recommended that people affected by emphysema take beta blockers. This scenario can be modeled by a AAF consisting of the three arguments reported above and the defeats $\delta_1 = (a, b)$, $\delta_2 = (b, a)$ and $\delta_3 = (c, b)$.

  It is easy to see that set $S = \{a, c\}$ is conflict-free and is an admissible extension, since the attack from b to a is counterattacked from both a and c. □

As a matter of fact, in the real world, arguments and defeats are often uncertain, thus, several proposals have been made to model uncertainty in AAFs, by considering weights, preferences, or probabilities associated with arguments and/or defeats. In this regard, [15, 30, 41, 40] have recently extended the original Dung framework in order to achieve probabilistic abstract argumentation frameworks (PrAFs), where uncertainty of arguments and defeats is modeled by exploiting probability theory. In particular, [30] proposed a PrAF where both arguments and defeats are associated with probability values and, in particular, they represent *independent probabilistic events*. Moreover, in [23] the complexity of the fundamental problem of computing the probability of extensions in this framework has been characterized.

However, in some cases it is not possible to assume that arguments are associated to probabilistic events independent from one another. For instance, this happens in the scenario described above, where arguments and defeats do not correspond to *independent probabilistic events*, as in the PrAF model defined in [30]. For instance, the probabilistic event associated with defeat $\delta_1 = (a, b)$ is the same as that associated with $\delta_2 = (b, a)$, meaning that δ_1 occurs iff δ_2 occurs (i.e., they are strongly correlated).

In order to deal with a scenario like that described in Example 1, arguments and defeats must be considered as correlated probabilistic events, as done in several approaches defined in the literature. In particular, in the approach defined in [41], instead of specifying arguments' and defeats' probabilities, users directly specify the unique probability distribution over the set of *possible worlds*, where a possible world is a set of probabilistic events corresponding to arguments and defeats. However, in this case, users may be required to specify a huge number of probability values (one for each possible world), as the number of possible worlds is exponential w.r.t. the number of arguments and defeats. Moreover, it can be the case that users are not aware of the probability value that should be assigned to a possible world, as it generally represents a complex scenario. Indeed, assigning probabilities to possible worlds is generally recognized to be so hard that in [27] it is shown that assigning probabilities to arguments and defeats is more intuitive and it is feasible in real-life scenarios.

Example 2. *Consider the AAF defined in Example 1. A corresponding PrAF will be characterized by the possible worlds [1] reported in Table 1. It is easy to see that devising the probability of each of the above-mentioned possible world is a hard task. Indeed, each possible world describes a complex scenario whose probability of occurrence is not easy to be estimated neither by an human expert nor by exploiting statistics.* □

An alternative way of representing the probability of each possible world is that of considering a set of independent probabilistic events (*basic events*) and characterize the occurrence of each argument and defeat as a *complex probabilistic event* specified by means

---

[1] Observe that combinations of arguments and defeats mentioning a defeat but not mentioning its arguments are not considered as they do not correspond to any realistic scenario.

PW-Name	Arguments	Defeats	PW-Name	Arguments	Defeats
pw_0	$\emptyset$	$\emptyset$	pw_{10}	a,c	$\emptyset$
pw_1	a	$\emptyset$	pw_{11}	a,b,c	$(a,b)(b,a),(c,b)$
pw_2	b	$\emptyset$	pw_{12}	a,b,c	$(a,b),(b,a)$
pw_3	c	$\emptyset$	pw_{13}	a,b,c	$(a,b),(c,b)$
pw_4	a,b	$(a,b),(b,a)$	pw_{14}	a,b,c	$(b,a),(c,b)$
pw_5	a,b	(a,b)	pw_{15}	a,b,c	(a,b)
pw_6	a,b	(b,a)	pw_{16}	a,b,c	(b,a)
pw_7	a,b	$\emptyset$	pw_{17}	a,b,c	(c,b)
pw_8	b,c	(c,b)	pw_{18}	a,b,c	$\emptyset$
pw_9	b,c	$\emptyset$			

Table 1: Possible worlds for any PrAF corresponding to the AAF of Example 1.

of an independence choice logic [37] formula on basic events as done in the *structured probabilistic argumentation frameworks* proposed in [23]. For instance, the occurrence of argument a of Example 1 can be written as the conjunction of the probabilistic events "John is affected by hypertension" and "Diuretics should be prescribed to people affected by hypertension", which clearly are independent from one another and whose marginal probability are sufficiently easy to be estimated.

Unfortunately, as proved in [23] the problem of computing the probability of extensions over structured probabilistic argumentation frameworks is $FP^{\#P}$-complete. Due to the complexity of the problem, it would seem that estimating the probability of extensions is the best thing to do, since computing it would require too much time. In this paper, we address the problem of computing/estimating the probability of extensions over structured probabilistic argumentation frameworks by devising a naive evaluation framework for *computing* the probability of extensions and a Monte-Carlo simulation algorithm for *estimating* it. We experimentally evaluate both the naive algorithm and the Monte-Carlo simulation one over two datasets, in order to establish whether some cases exist in which the naive algorithm can be profitably used instead of the Monte-Carlo one, and to identify cases in which the Monte-Carlo estimation algorithm represents the best choice.

Plan of the paper

In Section 2, we give an overview of Dung's abstract argumentation framework, and we define our structured probabilistic argumentation framework in Section 3. In Section 4, we define the naive algorithm for computing the probability of extensions and the Monte-Carlo simulation one for estimating it, and, in Section 5, we experimentally evaluate both algorithms. Finally, in Section 6 we discuss the related work and in Section 7 we draw conclusions.

2 Abstract Argumentation

In this section, we briefly overview Dung's abstract argumentation framework. An *abstract argumentation framework* [13] (*AAF*) is a pair $\langle A, D \rangle$, where A is a set whose elements are referred to as *arguments*, and $D \subseteq A \times A$ is a binary relation over A whose elements are referred to as *defeats* (or *attacks*). An argument is an abstract entity whose role is entirely determined by its relationships with other arguments. Given an AAF $\mathcal{A}$, we also refer to the set of its arguments and the set of its defeats as $Arg(\mathcal{A})$ and $Def(\mathcal{A})$, respectively. In the following we assume that $Arg(\mathcal{A})$ is finite, though Dung's original formulation did not require a finite set of arguments. We assume that a (finite) argumentation framework $\langle A, D \rangle$ is given and then it is the object of discourse, unless stated otherwise.

Given arguments $a, b \in A$, we say that a *defeats* b iff there is $(a, b) \in D$. Similarly, a set $S \subseteq A$ *defeats* an argument $b \in A$ iff there is $a \in S$ such that a *defeats* b.

A set $S \subseteq A$ of arguments is said to be *conflict-free* if there are no $a, b \in S$ such that a *defeats* b. An argument a is said to be *acceptable* w.r.t. $S \subseteq A$ iff $\forall b \in A$ such that b *defeats* a, there is $c \in S$ such that c *defeats* b.

Several semantics for AAFs have been proposed to identify "reasonable" sets of arguments, called *extensions*. We consider the following well-known semantics: *admissible* (`ad`), *stable* (`st`), *complete* (`co`), *grounded* (`gr`), *preferred* (`pr`) [13].
A set $S \subseteq A$ is said to be

- an *admissible extension* iff S is conflict-free and all its arguments are acceptable w.r.t. S;

- a *stable extension* iff S is conflict-free and S defeats each argument in $A \setminus S$;

- a *complete extension* iff S is admissible and S contains all the arguments that are acceptable w.r.t. S;

- a *grounded extension* iff S is a minimal (w.r.t. $\subseteq$) complete set of arguments;

- a *preferred extension* iff S is a maximal (w.r.t. $\subseteq$) complete set of arguments.

All the above-mentioned semantics, except the stable semantics, admit at least one extension, and the grounded admits exactly one extension [13, 14, 11, 43].

Example 3. *Consider the AAF $\langle A, D \rangle$, where the set A of arguments is $\{a, b, c\}$, and the set D of defeats is $\{\delta_1 = (c, a), \delta_2 = (c, b)\}$. As $S = \{c\}$ is conflict-free and c is acceptable w.r.t. S, it is the case that S is admissible. It is easy to see that $\emptyset$ is an admissible extension, whereas the sets $S' = \{a\}$ and $S'' = \{b\}$ are not admissible since S' (resp., S'') does not counterattack the attack from c to a (resp., b). Since $S = \{c\}$ is conflict-free and defeats both a and b, it is stable, complete, grounded, and preferred.* □

Example 4. *Consider an AAF $\langle A, D \rangle$, where $A = \{a, b, c, d\}$ and $D = \{(a, b), (b, c), (c, d), (d, a)\}$. The admissible sets are: $\emptyset$, $\{a, c\}$, and $\{b, d\}$. The empty set is a complete extension since no argument is acceptable in it, and both sets $\{a, c\}$ and $\{b, d\}$ are complete, preferred, and stable. The empty set is the grounded extension.* □

Given an AAF $\mathcal{A}$, a set $S \subseteq Arg(\mathcal{A})$ of arguments, and a semantics $sem \in \{\text{ad}, \text{st}, \text{co}, \text{gr}, \text{pr}\}$, we define function $ext(\mathcal{A}, sem, S)$ which returns *true* if S is an extension of $\mathcal{A}$ according to *sem*, *false* otherwise. For instance, for the AAF $\mathcal{A}$ in Example 4, $ext(\mathcal{A}, \text{ad}, \{a, c\})$ returns *true* since $\{a, c\}$ is an admissible set of arguments, whereas textitext$(\mathcal{A}, \text{ad}, \{a\})$ returns *false*.

3 Structured Probabilistic Argumentation Frameworks

Structured argumentation allows us to express correlations among events associated with arguments and defeats. Differently from other probabilistic abstract argumentation approaches, such as that proposed in [41], where expressing correlations between events associated to arguments and defeats requires specifying a probabilistic density function (PDF) at the level of possible worlds, our structured probabilistic argumentation framework proposal aims at implicitly specifying (in a compact way) those PDFs by exploiting the notions of *basic* and *complex probabilistic events*. As show in the following two examples, it is often possible to model events associated to arguments and defeats as complex probabilistic events which can be derived from basic probabilistic events that are independent from one another.

Example 5. *Continuing Example 1, we can express the probabilistic events associated with the arguments (reported again below for the sake of readability)*

   a: *Prescribe John diuretics since John has hypertension,*

   b: *Prescribe John beta blockers since John has hypertension,*

   c: *John has emphysema,*

and the defeats $\delta_1 = (a, b)$, $\delta_2 = (b, a)$ and $\delta_3 = (c, b)$, by introducing complex events based on the following basic events:

e_1: *John is affected by hypertension*,

e_2: *Beta blockers should be prescribed to people affected by hypertension*,

e_3: *Diuretics should be prescribed to people affected by hypertension*,

e_4: *John is affected by emphysema*,

e_5: *It is not recommended that people affected by emphysema take beta blockers*,

e_6: *It is not recommended to use both diuretics and beta blockers at the same time*. □

Given a set of basic independent events, we define complex events as follows.

Definition 1 (Complex events associated with arguments and defeats). *Let $\mathcal{E}$ be a set of basic probabilistic events. For each argument a, the complex event x_a associated with a is of the form $x_a = \phi$, where ϕ is a propositional formula over the basic events in $\mathcal{E}$. For each defeat $\delta = (a, b)$, the complex event x_δ associated with δ is of the form $x_\delta = \psi \wedge x_a \wedge x_b$, where ψ is a propositional formula over the basic events in $\mathcal{E}$, and x_a and x_b are the complex events associated with the arguments a and b of δ, respectively.*

Intuitively, the fact that a complex event x_δ associated with a defeat $\delta = (a, b)$ includes the conjunction of the complex events x_a and x_b means that the occurrence of the event x_δ is conditioned to the occurrence of both x_a and x_b. Note that every complex event associated with a defeat can be written by making explicit the complex events associated with the arguments involved in the defeat, that is, if $x_\delta = \psi \wedge x_a \wedge x_b$, $x_a = \phi$ and $x_b = \phi'$, we can write $x_\delta = \psi \wedge \phi \wedge \phi'$.

Example 6. *Using the basic events introduced in Example 5, it is easy to see that the definitions of the complex events associated with arguments a, b and c and with the defeats δ_1, δ_2 and δ_3 of Example 1 are as follows:*

- $x_a = e_1 \wedge e_3$

- $x_b = e_1 \wedge e_2$

- $x_c = e_4$

- $x_{\delta_1} = e_6 \wedge x_a \wedge x_b = e_6 \wedge e_1 \wedge e_3 \wedge e_2$

- $x_{\delta_2} = e_6 \wedge x_b \wedge x_a = e_6 \wedge e_1 \wedge e_2 \wedge e_3$

- $x_{\delta_3} = e_5 \wedge x_c \wedge x_b = e_5 \wedge e_4 \wedge e_1 \wedge e_2$ □

Thus the approach followed in our structured probabilistic argumentation framework is that of considering arguments and defeats associated with complex events that are in general not independent from one another. For instance, in the example above, x_a and x_b are correlated by the basic event e_1. However, every complex event is expressed as a propositional formula over basic independent events.

We now formally introduce the *structured probabilistic argumentation framework* and its semantics.

Definition 2 ($\mathcal{SF}$). *A structured probabilistic argumentation framework is a tuple $\mathcal{SF} = \langle A, D, \mathcal{E}, R, P_\mathcal{E} \rangle$ where*

- $\langle A, D \rangle$ *is an AAF,*

- $\mathcal{E}$ *is a set of basic independent probabilistic events,*

- R *is a set consisting of one complex probabilistic event for each argument in A and defeat in D, and*

- $P_\mathcal{E}$ *is a function assigning a probability value to each basic probabilistic event in $\mathcal{E}$.*

Example 7. *In examples 1, 5 and 6, we gradually introduced the structured probabilistic argumentation framework $\mathcal{SF} = \langle A, D, \mathcal{E}, R, P_\mathcal{E} \rangle$ where*

- $A = \{a, b, c\}$ *and* $D = \{\delta_1 = (a,b), \delta_2 = (b,a), \delta_3 = (c,b)\}$ *are the sets of arguments and defeats introduced in Example 1,*

- $\mathcal{E}$ *is the set of basic events of Example 5,*

- R *is the set of complex events of Example 6, and*

- $P_\mathcal{E}$ *is any probability values assignment for the basic probabilistic events in $\mathcal{E}$.*

To assign probabilities to basic events, we consider the confidence of the doctors in their own diagnoses and statistics about medical trials involving hypertensive patients and patients diagnosed with emphysema. In particular, assuming that doctors are 70% (resp., 30%) sure that John is hypertensive, we assign $P_\mathcal{E}(e_1) = 0.7$ (resp., $P_\mathcal{E}(e_4) = 0.3$). Furthermore, let's assume that medical trials statistics report that (i) the percentage of hypertensive patients which showed a significant health improvement after being treated with beta blockers (resp. diuretics) is the 90% (resp., 90%), (ii) that the percentage of patients diagnosed with emphysema which suffered some serious collateral effect after taking beta blockers is the 80%, and (iii) that the percentage of patients exhibiting some serious collateral effect after taking both beta blockers and diuretics is the 80%. Thus, we have that $P_\mathcal{E}(e_2) = 0.9$, $P_\mathcal{E}(e_3) = 0.9$, $P_\mathcal{E}(e_5) = 0.8$, and $P_\mathcal{E}(e_6) = 0.8$. □

It is worth noting that in Example 7 we needed to provide only 6 probability values (one for each basic event), while the number of probability values to be specified in Example 2 is 19 (one for each each possible scenario). Furthermore, the probability values given in Example 7 regard basic events, and are easier to be devised w.r.t. those to be attached at the complex scenarios represented by the possible worlds of Example 2.

The meaning of $\mathcal{SF} = \langle A, D, \mathcal{E}, R, P_\mathcal{E} \rangle$ is given in terms of its possible worlds, each of them representing a scenario that may occur in reality. Specifically, a possible world for $\mathcal{SF}$ is an AAF obtained using a subset of the arguments and defeats in A and D, respectively, and such that there is a subset of the basic events in $\mathcal{E}$ for which all and only the complex events associated with the arguments and defeats in the world hold.

Definition 3 (Possible world for structured probabilistic argumentation framework). *Given* $\mathcal{SF} = \langle A, D, \mathcal{E}, R, P_\mathcal{E} \rangle$, *a possible world for* $\mathcal{SF}$ *is an AAF* $\langle A', D' \rangle$ *such that*

(i) $A' \subseteq A$ *and* $D' \subseteq D \cap (A' \times A')$, *and*

(ii) *there is* $E \subseteq \mathcal{E}$ *such that*

  A) *all and only the arguments* $a \in A'$ *are such that the complex events* x_a *evaluate to true w.r.t.* E, *and*

  B) *all and only the defeats* $\delta \in D'$ *are such that the complex events* x_δ *evaluate to true w.r.t.* E.

In the following, we say that $E \subseteq \mathcal{E}$ supports world $w = \langle A', D' \rangle$, denoted as $E \models w$, iff Conditions (i) and (ii) of Definition 3 hold.

The set of the possible worlds of $\mathcal{SF}$ will be denoted as $pw(\mathcal{SF})$.

Example 8. *Consider the structured probabilistic argumentation framework* $\mathcal{SF}$ *of Example 7. It is easy to see that the AAF* $w_1 = \langle \{a, b\}, \{\delta_1, \delta_2\} \rangle$ *is a possible world for* $\mathcal{SF}$ *as Condition (i) trivially holds and Condition (ii) holds too, as there exists* $E = \{e_1, e_2, e_3, e_6\}$ *such that all and only the arguments and defeats in* w_1 *corresponds to events evaluating true w.r.t.* E. *Differently from the case of* w_1, *the AAF* $w_2 = \langle \{a, b\}, \{\delta_1\} \rangle$ *is not a possible world for* $\mathcal{SF}$ *as it is easy to see that for each* $E \subseteq \mathcal{E}$ *such that* x_{δ_1} *evaluates to true w.r.t.* E, *it is the case that* x_{δ_2} *evaluates to true w.r.t.* E *as well. Thus,* w_2 *is not a possible world for* $\mathcal{SF}$ *as it contains only* x_{δ_1}. □

Example 9. *The possible worlds of* $\mathcal{SF}$ *of Example 7 are reported in Table 2, where the right-most column shows a set of basic events supporting the world reported on the same row. It is worth noting that a given world can be supported by different sets of basic events. For instance, it can be easily checked that* w_1 *is also supported by* $\{e_1, e_4\}$ *and* $\{e_4, e_5\}$. □

Possible world w	Arguments	Defeats	A set $E \subseteq \mathcal{E}$ supporting w
w_0	$\emptyset$	$\emptyset$	$\emptyset$
w_1	c	$\emptyset$	e_4
w_2	a	$\emptyset$	e_1, e_3
w_3	a, c	$\emptyset$	e_1, e_3, e_4
w_4	b	$\emptyset$	e_1, e_2
w_5	b, c	$\emptyset$	e_1, e_2, e_4
w_6	b, c	δ_3	e_1, e_2, e_4, e_5
w_7	a, b	$\emptyset$	e_1, e_2, e_3
w_8	a, b	δ_1, δ_2	e_1, e_2, e_3, e_6
w_9	a, b, c	$\emptyset$	e_1, e_2, e_3, e_4
w_{10}	a, b, c	δ_1, δ_2	e_1, e_2, e_3, e_4, e_6
w_{11}	a, b, c	δ_3	e_1, e_2, e_3, e_4, e_5
w_{12}	a, b, c	$\delta_1, \delta_2, \delta_3$	$e_1, e_2, e_3, e_4, e_5, e_6$

Table 2: Possible worlds for the structured probabilistic argumentation framework of Example 7.

An interpretation π for $\mathcal{SF}$ is a probability distribution over the set $pw(\mathcal{SF})$ of the possible worlds, which is defined starting from the probability of basic events as follows.

As basic events are pairwise independent, the probability of a set $E \subseteq \mathcal{E}$ of basic events is as follows.

$$Pr(E) = \prod_{e \in E} P_{\mathcal{E}}(e) \cdot \prod_{e \in \mathcal{E} \setminus E} (1 - P_{\mathcal{E}}(e)). \tag{1}$$

That is, $Pr(E)$ is given by the product of the probabilities of the events belonging to E and the one's complements of the probabilities of the events that are in $\mathcal{E}$ but not in E. For instance, using the probability values of Example 7, $Pr(\{e_1, e_3\}) = P_{\mathcal{E}}(e_1) \cdot P_{\mathcal{E}}(e_3) \cdot (1 - P_{\mathcal{E}}(e_2)) \cdot (1 - P_{\mathcal{E}}(e_4)) \cdot (1 - P_{\mathcal{E}}(e_5)) \cdot (1 - P_{\mathcal{E}}(e_6)) = 0.7 \cdot 0.9 \cdot 0.1 \cdot 0.7 \cdot 0.2 \cdot 0.2 = 0.001764$.

Every possible world $w = \langle A', D' \rangle$ of $\mathcal{SF}$ is associated with probability $\pi(w)$ resulting from the sum of the probabilities $Pr(E)$ of the sets $E \subseteq \mathcal{E}$ supporting w, that is,

$$\pi(w) = \sum_{E \subseteq \mathcal{E} \wedge E \models w} Pr(E).$$

Example 10. *Consider the possible world $w_8 = \langle A', D' \rangle = \langle \{a, b\}, \{\delta_1, \delta_2\} \rangle$ shown in Table 2. As shown in Example 8, w_8 is supported by the set of events $E_1 = \{e_1, e_2, e_3, e_6\}$.*

It is easy to check that the only other set of events supporting w_8 is $E_2 = \{e_1, e_2, e_3, e_5, e_6\}$. Thus,
$$\pi(w_8) = Pr(E_1) + Pr(E_2) = 0.31752$$
where $Pr(E_1) = P_{\mathcal{E}}(e_1) \cdot P_{\mathcal{E}}(e_2) \cdot P_{\mathcal{E}}(e_3) \cdot (1 - P_{\mathcal{E}}(e_4)) \cdot (1 - P_{\mathcal{E}}(e_5)) \cdot P_{\mathcal{E}}(e_6) = 0.7 \cdot 0.9 \cdot 0.9 \cdot 0.7 \cdot 0.2 \cdot 0.8 = 0.063504$, and
$Pr(E_2) = P_{\mathcal{E}}(e_1) \cdot P_{\mathcal{E}}(e_2) \cdot P_{\mathcal{E}}(e_3) \cdot (1 - P_{\mathcal{E}}(e_4)) \cdot P_{\mathcal{E}}(e_5) \cdot P_{\mathcal{E}}(e_6) = 0.7 \cdot 0.9 \cdot 0.9 \cdot 0.7 \cdot 0.8 \cdot 0.8 = 0.254016$. □

The probability $Pr_{\mathcal{SF}}^{sem}(S)$ that a set S of arguments is an extension according to a given semantics *sem* is defined as the sum of the probabilities $\pi(w)$ of the possible worlds w where S is an extension according to *sem*, that is, the sum of the probabilities of the possible worlds w for which it holds that *ext(w, sem, S)* = true [2].

Definition 4 ($Pr_{\mathcal{SF}}^{sem}(S)$). *Given a structured probabilistic argumentation framework $\mathcal{SF} = \langle A, D, \mathcal{E}, R, P_{\mathcal{E}} \rangle$, a set $S \subseteq A$ of arguments, and a semantics sem, the probability $Pr_{\mathcal{SF}}^{sem}(S)$ that S is an extension according to sem is as follows*

$$Pr_{\mathcal{SF}}^{sem}(S) = \sum_{\substack{w \in pw(\mathcal{SF}) \\ \wedge ext(w, sem, S)}} \pi(w) = \sum_{\substack{w \in pw(\mathcal{SF}) \\ \wedge ext(w, sem, S) \\ \wedge E \subseteq \mathcal{E} \wedge E \models w}} Pr(E).$$

Example 11. *To compute the probability $Pr_{\mathcal{SF}}^{pr}(\{b\})$ that the set $\{b\}$ is a preferred extension, we first identify the possible worlds w of Table 2 such that $\{b\}$ is a preferred extension in w (that is, such that ext(w, pr, $\{b\}$)=true). It is easy to check that these worlds are w_4 and w_8. For any other world in Table 2, function ext(w, pr, $\{b\}$)=false. Indeed, for worlds w_0, w_1, w_2, w_3, ext(w, pr, $\{b\}$) evaluates to false since these worlds do not contain argument b; for worlds w_6, w_{11}, w_{12}, ext(w, pr, $\{b\}$)=false since these worlds contains the defeat $\delta_3 = (c, b)$; finally, for worlds w_5, w_7, w_9, ext(w, pr, $\{b\}$)=false since the set $\{b\}$ is not a maximal complete extension (argument a or c, or both, are acceptable w.r.t. $\{b\}$). Therefore,*

$$Pr_{\mathcal{SF}}^{pr}(\{b\}) = \pi(w_4) + \pi(w_8),$$

where $\pi(w_8)$ is as shown in Example 10 and $\pi(w_4)$ is the sum of the probability of the sets of events supporting w_4, that is $Pr(E_1) + Pr(E_2) + Pr(E_3) + Pr(E_4)$, where $E_1 = \{e_1, e_2\}$, $E_2 = \{e_1, e_2, e_5\}$, $E_3 = \{e_1, e_2, e_6\}$, and $E_4 = \{e_1, e_2, e_5, e_6\}$. □

We report below a result concerning the computational complexity of computing $Pr_{\mathcal{SF}}^{sem}(\{b\})$. Let $\text{PROB}_{\mathcal{SF}}^{sem}(S)$ denote the problem of computing the probability $Pr_{\mathcal{SF}}^{sem}(S)$. It was shown in [23] that, for several popular semantics, $\text{PROB}_{\mathcal{SF}}^{sem}(S)$ is complete for the

---
[2] If $S \not\subseteq Arg(w)$ then *ext(w, sem, S)*=false for every semantics *sem*.

complexity class $FP^{\#P}$, that is, the class of functions computable by a polynomial-time Turing machine with a $\#P$ oracle, where $\#P$ is the complexity class of the functions f such that f counts the number of accepting paths of a nondeterministic polynomial-time Turing machine [42].

Fact 1. $\text{PROB}_{\mathcal{SF}}^{sem}(S)$, *where* sem $\in \{$ad, st, co, gr, pr, $\}$, *is* $FP^{\#P}$*-complete* [3].

In the next section, we address the problem of computing or estimating the probability $Pr_{\mathcal{SF}}^{sem}(S)$ that a set of argument S is an extension of a structured probabilistic argumentation framework according to semantics sem $\in \{$ad, st, co, gr, pr$\}$.

4 Computing or Estimating Extensions' Probabilities

In this section we present a naive evaluation framework for computing $Pr_{\mathcal{SF}}^{sem}(S)$ and a Monte-Carlo simulation algorithm for estimating $Pr_{\mathcal{SF}}^{sem}(S)$.

4.1 Computing $Pr_{\mathcal{SF}}^{sem}(S)$

We now introduce the exhaustive approach for computing $Pr_{\mathcal{SF}}^{sem}(S)$.

Algorithm 1. *Computing* $Pr_{\mathcal{SF}}^{sem}(S)$
Input: $\mathcal{SF} = \langle A, D, \mathcal{E}, R, P_{\mathcal{E}} \rangle$; $S \subseteq A$; sem;
Output: $Pr_{\mathcal{SF}}^{sem}(S)$
1: $Pr_S = 0$;
2: $SE = computeSubsets(\mathcal{E})$;
3: **for each** $E \in SE$
4:     $Pr_E = 1$;
5:     **for each** $e \in E$
6:         $Pr_E = Pr_E \cdot P_{\mathcal{E}}(e)$;
7:     **for each** $e \in \mathcal{E} \setminus E$
8:         $Pr_E = Pr_E \cdot (1 - P_{\mathcal{E}}(e))$;
9:     $Arg = \emptyset, Def = \emptyset$;
10:    **for each** $a \in A$ such that $x_a = \phi \in R$
11:        **if** (ϕ evaluates to true w.r.t E)
12:            $Arg = Arg \cup \{a\}$;
13:    **for each** $\delta = (a, b) \in D$ such that $a, b \in Arg$ and $x_\delta = \psi \wedge x_a \wedge x_b \in R$
14:        **if** (ψ evaluates to true w.r.t E)

---
[3] In [23], the $FP^{\#P}$-completeness of $\text{PROB}_{\mathcal{SF}}^{sem}(S)$ was also shown for the *ideal* [14], *semi-stable* [11], and *stage* [43] semantics.

15:     $Def = Def \cup \{\delta\}$;
16:   **if** $ext(\langle Arg, Def \rangle, sem, S)$
17:     $Pr_S = Pr_S + Pr_E$;
18: **return** Pr_S

In brief, given a structured probabilistic argumentation framework $\mathcal{SF} = \langle A, D, \mathcal{E}, R, P_{\mathcal{E}} \rangle$, a set S, and a semantic sem, Algorithm 1 computes $Pr_{\mathcal{SF}}^{sem}(S)$ by (i) generating all the sets E of basic events obtainable from $\mathcal{E}$, (ii) building all the possible worlds of $\mathcal{SF}$ from the sets E, and (iii) summing the probabilities of the possible worlds in which S is an extension according to sem. In more detail, at line 2, Algorithm 1 builds all the subsets of $\mathcal{E}$ by calling function *computeSubsets*. Next, for each subset E, Algorithm 1 computes the probability of E as defined in Equation 1 (lines $4-8$), and builds the possible world $\langle Arg, Def \rangle$ supporting E (lines $9-15$). Specifically, the possible world $\langle Arg, Def \rangle$ is built by adding to Arg all the arguments $a \in A$ such that the propositional formula of its corresponding event x_a is true w.r.t. E, and all the defeats (a, b) such that both a and b belong to Arg and the propositional formula of the corresponding event x_δ is true w.r.t. E. At line 16, function *ext* is called to verify whether S is an extension according to sem in $\langle Arg, Def \rangle$: if *ext* evaluates to true, variable Pr_S is incremented by the probability of E, according to Definition 4. At the end of the outermost loop, the value of Pr_S is returned as output.

4.2 Estimating $Pr_{\mathcal{SF}}^{sem}(S)$

We now introduce the Monte-Carlo approach for estimating $Pr_{\mathcal{SF}}^{sem}(S)$.

Algorithm 2. *Estimating* $Pr_{\mathcal{SF}}^{sem}(S)$
Input: $\mathcal{SF} = \langle A, D, \mathcal{E}, R, P_{\mathcal{E}} \rangle$; $S \subseteq A$; sem; An error level ϵ; A confidence level $1 - \alpha$
Output: $\widehat{Pr}_{\mathcal{SF}}^{sem}(S)$ s.t. $Pr_{\mathcal{SF}}^{sem}(S) \in [\widehat{Pr}_{\mathcal{SF}}^{sem}(S) - \epsilon, \widehat{Pr}_{\mathcal{SF}}^{sem}(S) + \epsilon]$ with confidence $1 - \alpha$
1: $x = n = 0$;
2: **do**
3:     $E = \emptyset$, $Arg = \emptyset$, $Def = \emptyset$;
4:     **for each** $e \in \mathcal{E}$
5:         generate a number r in $[0, 1)$;
6:         **if** $(r \leq P_{\mathcal{E}}(e))$
7:             $E = E \cup \{e\}$;
8:     **for each** $a \in A$ such that $x_a = \phi \in R$
9:         **if** (ϕ evaluates to true w.r.t E)
10:            $Arg = Arg \cup \{a\}$;
11:    **for each** $\delta = (a, b) \in D$ such that $a, b \in Arg$ and $x_\delta = \psi \wedge x_a \wedge x_b \in R$
12:        **if** (ψ evaluates to true w.r.t E)

13:     $Def = Def \cup \{\delta\}$;
14:   **if** $ext(\langle Arg, Def \rangle, sem, S)$
15:     $x=x+1$;
16:   $n=n+1;\ p = \frac{x+z_{1-\alpha/2}^2/2}{n+z_{1-\alpha/2}^2};\ n' = \frac{z_{1-\alpha/2}^2 \cdot p \cdot (1-p)}{\epsilon^2} - z_{1-\alpha/2}^2$
17: **while** $n \leq n'$
18: **return** x/n

Given a structured probabilistic argumentation framework $\mathcal{SF} = \langle A, D, \mathcal{E}, R, P_{\mathcal{E}} \rangle$, a set S, a semantic sem, a confidence level $1 - \alpha$, and an error level ϵ, the Monte-Carlo estimation algorithm consists of (i) generating a number n of sets $E \subseteq \mathcal{E}$ of basic independent probabilistic events, each of them corresponding to a possible world w, (ii) checking if S is extension according to sem in the generated w, (iii) returning as output the number x/n, where x is the number of possible worlds wherein S is an extension according to sem. Specifically, given a structured probabilistic argumentation framework $\mathcal{SF}$, a set S, a semantic sem, an error level ϵ, and a confidence level $1 - \alpha$, the algorithm returns an estimate $\widehat{Pr}_{\mathcal{SF}}^{sem}(S)$ of $Pr_{\mathcal{SF}}^{sem}(S)$ such that $Pr_{\mathcal{SF}}^{sem}(S)$ lies in the interval $\widehat{Pr}_{\mathcal{SF}}^{sem}(S) \pm \epsilon$ with a confidence level $1 - \alpha$. The number n of sets of basic independent probabilistic events to be sampled to achieve the required error level ϵ with confidence level $1 - \alpha$ is determined by exploiting the Agresti-Coull interval [1]. In particular, according to [1], the estimated value p of $Pr_{\mathcal{SF}}^{sem}(S)$ after x successes in n samples is $p = \frac{x+(z_{1-\alpha/2}^2)/2}{n+z_{1-\alpha/2}^2}$, where $z_{1-\alpha/2}$ is the $1 - \alpha/2$ quantile of the normal distribution, and the number of samples ensuring that the error level is ϵ with confidence level $1 - \alpha$ is $n = \frac{z_{1-\alpha/2}^2 \cdot p \cdot (1-p)}{\epsilon^2} - z_{1-\alpha/2}^2$.

Each iteration of Algorithm 2 consists of the following steps. First, Algorithm 2 generates a set E of basic probabilistic events (lines 4-7), by adding to E each basic event (belonging to $\mathcal{E}$) whose probability is greater than or equal to a randomly generated number. Next, Algorithm 2 builds the possible world $w = \langle Arg, Def \rangle$ such that E supports w (lines 8-13) as follows. First, all the arguments of A whose propositional formula evaluates to true w.r.t. E are added to Arg, and, next, all the defeats (a, b) such that both a and b belong to Arg and whose propositional formula evaluates to true w.r.t. E are added to Def. After generating a possible world w, Algorithm 2 checks if S is an extension according to sem in w, and, if this is the case, it increments x's value. After that, it computes the number n' of samples to be generated according to the Agresti-Coull interval. Finally, Algorithm 2 returns x/n as output. As stated below, Algorithm 2 is sound.

Proposition 1. Let $\mathcal{SF} = \langle A, D, \mathcal{E}, R, P_{\mathcal{E}} \rangle$, and $S \subseteq A$. Let ϵ be an error level, and $1 - \alpha$ a confidence level. The estimate $\widehat{Pr}_{\mathcal{SF}}^{sem}(S)$ returned by Algorithm 2 is such that $Pr_{\mathcal{SF}}^{sem}(S) \in [\widehat{Pr}_{\mathcal{SF}}^{sem}(S) - \epsilon,\ \widehat{Pr}_{\mathcal{SF}}^{sem}(S) + \epsilon]$ with confidence level $1 - \alpha$.

The proof easily follows from the definition of the Agresti-Coull interval, which is applicable to our case since the underlying distribution is binomial: in fact, the probability of success for a set (i.e., the probability that S is an extension according to *sem* in the possible world corresponding to the generated set of basic probabilistic events) does not influence the probability of success for the other sets picked by the sampler.

5 Experiments

The experimental study described in this section is aimed at assessing the efficiency of the naive approach compared to the Monte-Carlo simulation algorithm. All the experiments were run on an Intel(R) Core(TM) i5 CPU M520, 2.40GHz. We experimentally evaluate both the naive and the Monte-Carlo algorithm, in order to establish whether some cases exist in which the naive algorithm can be profitably used instead of the Monte-Carlo one, and to identify other cases in which the Monte-Carlo estimation algorithm represents the best choice.

5.1 Data sets

In our experimental studies we consider the two different datasets *standard* (named $DS1$) and *borderline* (named $DS2$): the former contains input instances (of the form $\langle \mathcal{SF}, S, sem \rangle$) for which $Pr_{\mathcal{SF}}^{sem}(S)$ is different from 0 and 1, while the latter consists of input instances for which $Pr_{\mathcal{SF}}^{sem}(S)$ is equal to 0 or 1. Splitting the input instances into these two datasets allows us to perform a fair comparison between Algorithm 1 and Algorithm 2, as the latter is likely to run very fast over input instances for which $Pr_{\mathcal{SF}}^{sem}(S)$ is very close to 0 or 1, due to the intrinsic features of the Monte-Carlo approach. Hence, $DS2$ can be seen as an easy test case for Algorithm 2.

Each data set consists of three subsets of structured probabilistic argumentation frameworks which differ in the number of arguments. Specifically, the first subset contains the frameworks consisting of 5 arguments, the second those consisting of 10 arguments and the third those having 15 arguments. Each subset was obtained by varying the number of basic events over the set $\{15, 20, 25, 30, 35, 40, 45\}$ and by generating 10 frameworks for each element in the set. Therefore each subset contains 70 frameworks, which means that both $DS1$ and $DS2$ contain 210 frameworks.

5.2 Results and discussion

In this section, we evaluate the efficiency of Algorithm 1, which computes $Pr_{\mathcal{SF}}^{sem}(S)$, and Algorithm 2, which estimates $Pr_{\mathcal{SF}}^{sem}(S)$, with $sem \in \{\text{ad}, \text{co}, \text{gr}, \text{pr}, \text{st}\}$. All the runs of the Monte-Carlo simulation algorithm were done by setting the confidence level to 95%

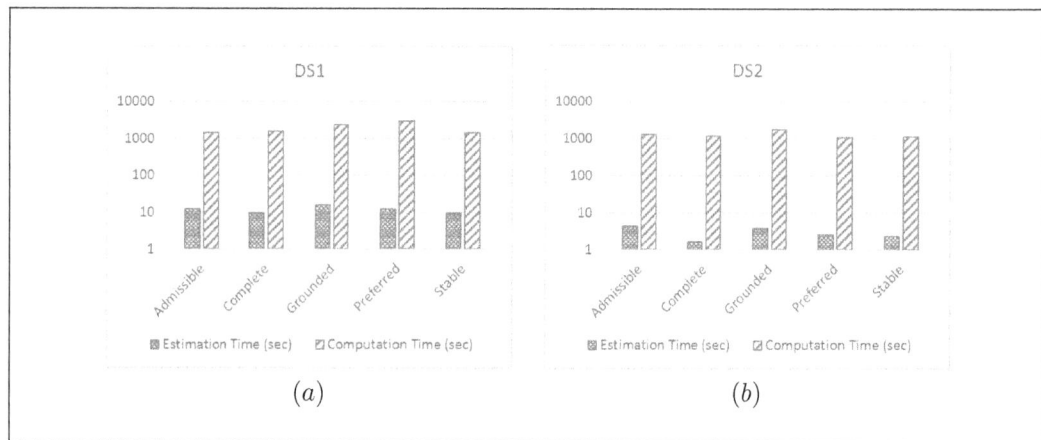

Figure 1: Execution time vs. semantics over $DS1$ (a) and $DS2$ (b)

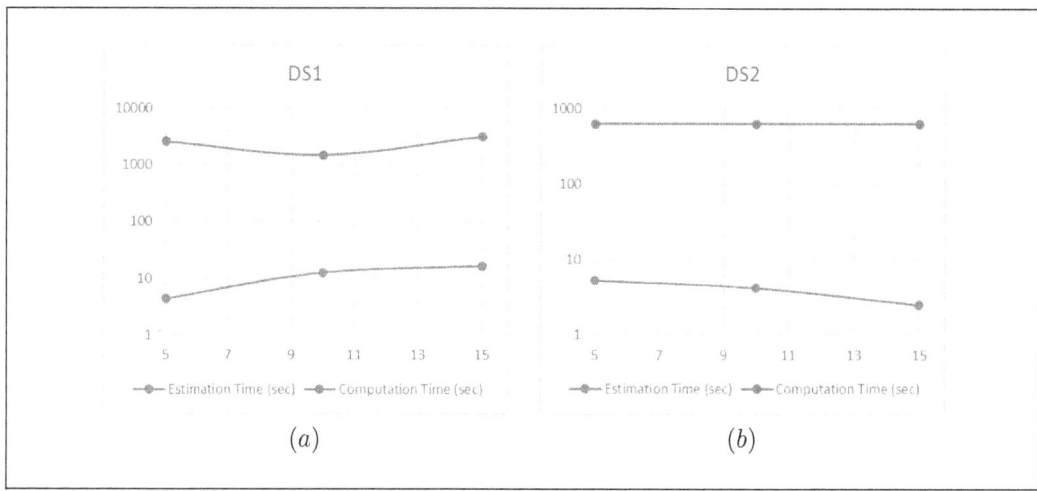

Figure 2: Execution time vs. number of arguments over $DS1$ (a) and $DS2$ (b)

and the error level to 0.05%. Figure 1 reports both the estimation and computation times versus the different semantics over $DS1$ (a) and $DS2$ (b), Figure 2 reports both the estimation and computation times versus the number of arguments over $DS1$ (a) and $DS2$ (b), and Figure 3 reports both the estimation and computation times versus the number of basic events over $DS1$ (a) and $DS2$ (b). Note that y axes are in log scale.

From the experimental evaluation, it turns out that the execution times of both computing and estimating $Pr_{\mathcal{S}\mathcal{F}}^{sem}(S)$ do not depend on the semantics and the number of arguments. Moreover, it turns out that the time required by Algorithm 1 grows exponentially with the number of basic events, while the time required by Algorithm 2 is independent from the

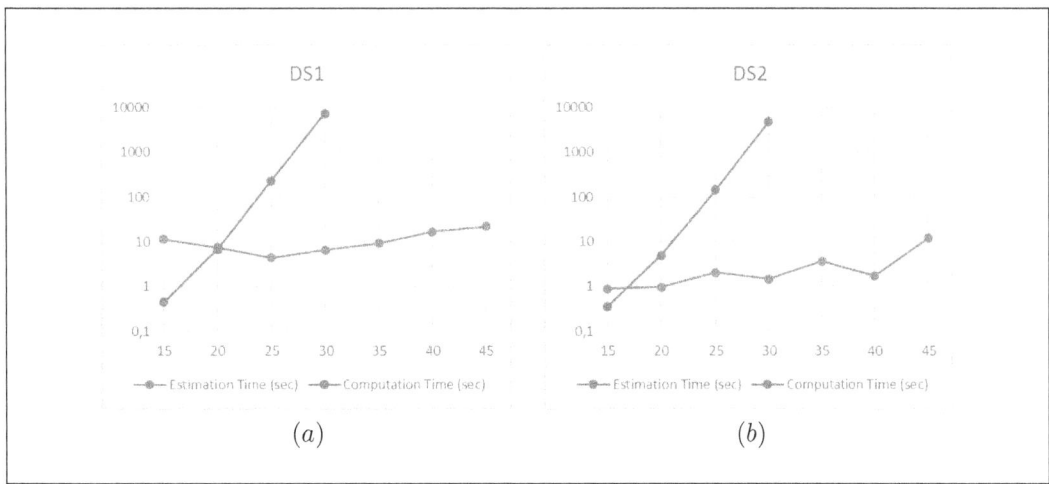

Figure 3: Execution time vs. number of basic events over $DS1$ (a) and $DS2$ (b)

number of basic events. Finally, not surprisingly, it turns out that the estimating $Pr^{sem}_{\mathcal{SF}}(S)$ is faster on $DS2$ than on $DS1$.

Overall, we can draw the conclusion that computing $Pr^{sem}_{\mathcal{SF}}(S)$ is efficient enough when the number of basic events is less than or equal to 20, while in the case that the number of basic events is larger (greater than 25) it is practically unfeasible, as on average its running time is of more than one hour and a half. Hence, Algorithm 1 should be used in the case that the number of basic events is less than or equal to 20, as in this case computing probabilities is easier than estimating them, and conversely Algorithm 2 should be used in the other cases.

6 Related work

Recently, approaches for handling uncertainty in AAFs by relying on probability theory have been proposed in [15, 30, 40, 41]. Specifically, with the aim of modeling jury-based dispute resolutions, [15] proposed a probabilistic abstract argumentation framework (PrAF) where uncertainty is taken into account by specifying probability distribution functions (PDFs) over possible worlds and it shown how an instance of the proposed PrAF can be obtained by specifying a probabilistic assumption-based argumentation framework (introduced by themselves). Differently from the previous approach, [30] proposed a PrAF where probabilities are directly associated with arguments and defeats, instead of being associated with possible worlds. [30] claimed that computing the probability $Pr(S)$ that a set S of arguments belongs to an extension requires exponential time for every semantics, and then proposed a Monte-Carlo simulation approach to approximate $Pr(S)$. However, in [22, 23] it was shown that the usage of approximation is more appropriate for those semantics *sem* for which $\text{PROB}_{\mathcal{F}}^{sem}(S)$ is hard, while for the tractable semantics (admissible and stable) the exact solution of $\text{PROB}_{\mathcal{F}}^{sem}(S)$ can be found in polynomial-time. Recently, in [21] it was devised an optimized Monte-Carlo simulation approach which is able to estimate $Pr^{sem}(S)$, with $sem \in \{\text{co}, \text{gr}, \text{pr}\}$, using much fewer samples than the original approach proposed in [30], resulting in a significantly more efficient estimation technique. Specifically, the proposed approach exploits the tractability results presented in [22, 23] for estimating $Pr^{sem}(S)$ as $Pr^{sem|AD}(S) \times Pr^{\text{ad}}(S)$ (where $Pr^{sem|AD}(S)$ is the conditional probability that S is an extension according to *sem* given that S is an admissible extension), given that $Pr^{\text{ad}}(S)$ can be computed efficiently. We point out that, the probabilistic argumentation framework (PrAF) introduced in [30], and investigated in [22, 21, 23, 25] can be expressed by means of a structured probabilistic argumentation framework where ϕ and ψ in Definition 1 are basic events. That is, arguments are associated with basic events, and the complex event x_δ associated with a defeat $\delta = (a, b)$ is simply of the form $x_\delta = \psi \wedge x_a \wedge x_b$ where ψ is a basic event (this means that the occurrence of the event x_δ is conditioned to the occurrence of both x_a and x_b).

In [30], as well as in [15, 40] and [22, 21, 23, 25, 24], $Pr^{sem}(S)$ is defined as the sum of the probabilities of the possible worlds where S is an extension according to semantics *sem*. [41] instead did not define a probabilistic version of a classical semantics, but introduced a new probabilistic semantics. This semantics is based on *p-justifiable* PDFs defined over the set of possible worlds: given an AAF $\mathcal{A} = \langle A, D \rangle$, a PDF f is a function assigning a probability to each possible world w of $\mathcal{A}$ [4], and it is said to be p-justifiable iff for each argument $a \in A$, it holds that (*i*) for each argument b defeating a, the probability that a is in an extension according to f is lower than or equal to the one's complement of the probability that b is in an extension according to f; (*ii*) the probability that a is in an extension according to f is greater than or equal to the one's complement of the sum, over arguments b defeating a, of the probability that b is in an extension according to p.

In the above-cited works probability theory is recognized as a fundamental tool to model uncertainty. In this regard, a deeper understanding of the role of probability theory in abstract argumentation was developed in [26, 27], where the *justification* and the *premise* perspectives of probabilities of arguments are introduced. According to the former perspective the probability of an argument indicates the probability that it is justified in appearing in the argumentation system. In contrast, the premise perspective views the probability of an argument as the probability that the argument is true based on the degrees to which the premises supporting the argument are believed to be true. Starting from these perspectives, in [27], a formal framework showing the connection among argumentation theory, classical

---

[4]Observe that any subset of A is considered as a possible world in [41], since defeat $(a, b) \in D$ occurs if and only if both a and b occur.

logic, and probability theory was investigated. Furthermore, qualification of attacks is addressed in [28], where an investigation of the meaning of the uncertainty concerning defeats in probabilistic abstract argumentation is provided.

We now discuss works [31, 40] that are closer to the structured probabilistic argumentation framework addressed in this paper. Like in our framework, in these works, independence assumption between events associated with arguments/defeats is used only in a very limited form. In particular, an interesting step towards the integration of argumentation frameworks and probability theory has been recently achieved by [31], where the evidential argumentation framework (EAF) [35] has been extended by assigning probabilities to the items of support relations. This way, probabilistic evidential argumentation frameworks (PrEAFs) are obtained, which model inter-argument dependencies by assigning conditional probabilities between arguments and their supporting arguments. As pointed out in [31], PrEAFs can be viewed as a generalization of PrAFs where the assumption that arguments are mutually independent is relaxed. However, the items of the support relation are assumed to be independent from one another, which makes the approach in [31] similar in spirit to our structured probabilistic argumentation framework, where basic events are independent from one another. In our framework, dependency between arguments can be expressed by complex events in terms of FO propositional formulas of basic events. For instance, probabilistic support of an argument a by argument b can be expressed using the complex event $x_a = x_b \wedge x_s$ where x_b is the complex event of b and x_s is the probabilistic event associated with the support (this can be generalized to the case of support provided by several arguments). In turns, x_b could be supported by other arguments, and this can be recursively captured by its complex event. The fact that in [31] EAFs induced by a given PrEAF include only supported arguments is captured by our definition of possible world: in fact, only arguments such that their complex events (including support events) evaluate to true can belong to a possible world. However, since complex events can be defined using general FO formulas over a given set of basic events, there are dependencies among arguments/defeats that can be expressed in our structured probabilistic argumentation framework and cannot be expressed in PrEAFs, such as $x_a = (x_b \vee x_c) \wedge \neg(x_b \wedge x_c)$, saying that argument a occurs iff either argument b or c occurs. This suggests that our algorithms could be used to compute the probability of reasonable sets of arguments in PrEAFs, but also that the simpler structure of dependencies in PrEAFs could be exploited to modify the construction of possible worlds by exploiting the strategy proposed in [31] for finding inducible EAFs. In [29] Probabilistic Extended Evidential Argumentation Frameworks (PrEEAFs) have been introduced, which extend PrEAFs with the possibility of associating probabilities to defeats and introduce further constructs such as defeats to defeats. As shown in [29], every PrEEAF can be translated into an equivalent PrEAF. This fact, in turn, entails that our structured probabilistic argumentation frameworks are more expressive than PrEEAFs. ASPIC framework [38], a general abstract model of argumentation with structured arguments that has as a special case the *assumption-based argumentation* (*ABA*) [16], has been recently extended in [40], where an instantiation of PrAFs with a probabilistic version of a general fragment of ASPIC, called *p-ASPIC*, has been introduced. *p-ASPIC* allows the use of strict rules, defeasible rules, rebutting attacks and undercutting attacks in a probabilistic context. The approach adopted in [40] for translating their PrAFs into p-ASPIC relies on assigning probabilities to the ASPIC rules and interpreting the resulting probabilistic rules as basic events independent one from another (while arguments defined by using these rules are in general not independent, as it happens for our complex events). It is worth noting that p-ASPIC models can be simulated using the structured probabilistic argumentation frameworks proposed in this paper. This can be done by basically replacing a p-ASPIC rule of the form $\phi \leftarrow_p \psi$ with an argument a whose complex event is $x_a = (\phi \leftarrow \psi) \wedge e$, where e is a basic event whose probability is p. Hence, the algorithms proposed for computing or estimating the probability of extensions over a structured probabilistic argumentation framework can be easily adapted for computing or estimating the probability of extensions over a p-ASPIC instance.

Besides the approaches that model uncertainty in AAFs by relying on probability theory, many proposals have been made where uncertainty is represented by exploiting weights or preferences on arguments and/or defeats [8, 6, 4, 33, 18, 12]. In [8] each argument is associated with a numeric value, and a set of possible orders (preferences) among the values is defined. Here, a defeat succeeds w.r.t. a specific value order only if the value associated with the defeated argument is not preferred to the value associated with the defeating argument in the value order. All the semantics are extended to take into account this notion of defeat. [6] extends [8] by introducing preferences among sets of arguments, exploiting the values associated with the arguments. The aim is that of choosing the best set of arguments among those satisfying a (classic) semantics. In [33] arguments can express preferences between other arguments, determining whether defeats succeed or not, while in [4] a defeat succeeds only if the defeated is not preferred to the attacker, on the basis of a preference relation between arguments. [32] introduces preferences between defeats, with the aim of finding the extensions providing best defenses for their elements. [18] associates attacks with weights, and proposes new semantics extending the classical ones on the basis of a threshold β. Specifically, a set S of arguments is a β-*sem* extension, where *sem* is a semantics, iff S is an extension according *sem* in the AAF obtained by removing, from the original set of defeats, a subset of defeats whose weights sum up at most to β. For a semantics *sem*, a set S of arguments is preferred to another set S' iff S requires a smaller value of β to be a *sem* extension. A study of the computational complexity of the problem of computing the proposed semantics is carried out. [12] extends [18] by considering other aggregation functions over weights apart from sum.

In addition to the above-mentioned approaches, another interesting approach to represent uncertainty in argumentation is that based on using possibility theory, such as done in [5, 2, 3, 34]. In particular, [34] proposed an argumentation-based possibilistic decision making framework which, though able to capture uncertain information and exceptions/defaults, has the nice property that argument inference is polynomial-time computable (this feature follows from the fact that the framework is based on non-monotonic inference of Possibilistic Well-Founded Semantics which is tractable [36]).

Although the approaches based on weights, preferences, possibilities, or probabilities to model uncertainty have been proved to be effective in different contexts, there is no common agreement on what kind of approach should be used in general. In this regard, [26, 27] observed that the probability-based approaches may take advantage from relying on a well-established and well-founded theory, whereas the approaches based on weights or preferences do not conform to well-established theories yet.

7 Conclusions and future work

In this paper, we addressed the problem of computing/estimating the probability of extensions over structured probabilistic argumentation frameworks, by devising a naive algorithm for computing the probability of extensions and a Monte-Carlo simulation algorithm for estimating it. We experimentally evaluated both the naive algorithm and the Monte-Carlo one over two datasets, and found some cases for which computing the probability results to be more convenient than estimating it, despite the high complexity of the problem.

An interesting direction for future work is that of identifying tractable cases of the problem of computing the probability of extensions over structured probabilistic argumentation frameworks, for instance by considering restricted forms of propositional boolean formulas for expressing complex events.

References

[1] Alan Agresti and Brent A. Coull. Approximate is better than "exact" for interval estimation of binomial proportions. *The American Statistician*, 52(2):119–126, 1998.

[2] Teresa Alsinet, Carlos Iván Chesñevar, Lluis Godo, Sandra Sandri, and Guillermo Ricardo Simari. Formalizing argumentative reasoning in a possibilistic logic programming setting with fuzzy unification. *International Journal of Approximate Reasoning*, 48(3), 2008.

[3] Teresa Alsinet, Carlos Iván Chesñevar, Lluis Godo, and Guillermo Ricardo Simari. A logic programming framework for possibilistic argumentation: Formalization and logical properties. *Fuzzy Sets and Systems*, 159(10):1208–1228, 2008.

[4] Leila Amgoud and Claudette Cayrol. A reasoning model based on the production of acceptable arguments. *Annals of Mathematics and Artificial Intelligence*, 34(1-3):197–215, 2002.

[5] Leila Amgoud and Henri Prade. Reaching agreement through argumentation: A possibilistic approach. In *Proc. of International Conference on Principles of Knowledge Representation and Reasoning (KR)*, pages 175–182, 2004.

[6] Leila Amgoud and Srdjan Vesic. A new approach for preference-based argumentation frameworks. *Annals of Mathematics and Artificial Intelligence*, 63(2):149–183, 2011.

[7] Pietro Baroni and Massimiliano Giacomin. Semantics of abstract argument systems. In *Argumentation in Artificial Intelligence*, pages 25–44. Springer US, 2009.

[8] Trevor J. M. Bench-Capon. Persuasion in practical argument using value-based argumentation frameworks. *Journal of Logic and Computation*, 13(3):429–448, 2003.

[9] Trevor J. M. Bench-Capon and Paul E. Dunne. Argumentation in artificial intelligence. *Artificial Intelligence*, 171(10-15):619–641, 2007.

[10] Philippe Besnard and Anthony Hunter, editors. *Elements Of Argumentation*. The MIT Press, 2008.

[11] Martin Caminada. Semi-stable semantics. In *Proc. of International Conference on Computational Models of Argument (COMMA)*, pages 121–130, 2006.

[12] Sylvie Coste-Marquis, Sébastien Konieczny, Pierre Marquis, and Mohand Akli Ouali. Weighted attacks in argumentation frameworks. In *Proc. of International Conference on Principles of Knowledge Representation and Reasoning (KR)*, 2012.

[13] Phan Minh Dung. On the acceptability of arguments and its fundamental role in nonmonotonic reasoning, logic programming and n-person games. *Artificial Intelligence*, 77(2):321–358, 1995.

[14] Phan Minh Dung, Paolo Mancarella, and Francesca Toni. Computing ideal sceptical argumentation. *Artificial Intelligence*, 171(10-15):642–674, 2007.

[15] Phan Minh Dung and Phan Minh Thang. Towards (probabilistic) argumentation for jury-based dispute resolution. In *Proc. of International Conference on Computational Models of Argument (COMMA)*, pages 171–182, 2010.

[16] P.M. Dung, R.A. Kowalski, and F. Toni. Assumption-based argumentation. In *Argumentation in Artificial Intelligence*, pages 199–218. Springer US, 2009.

[17] Paul E. Dunne. The computational complexity of ideal semantics. *Artificial Intelligence*, 173(18):1559–1591, 2009.

[18] Paul E. Dunne, Anthony Hunter, Peter McBurney, Simon Parsons, and Michael Wooldridge. Weighted argument systems: Basic definitions, algorithms, and complexity results. *Artificial Intelligence*, 175(2):457–486, 2011.

[19] Paul E. Dunne and Michael Wooldridge. Complexity of abstract argumentation. In *Argumentation in Artificial Intelligence*, pages 85–104. Springer US, 2009.

[20] Wolfgang Dvořák and Stefan Woltran. Complexity of semi-stable and stage semantics in argumentation frameworks. *Information Processing Letters*, 110(11):425–430, 2010.

[21] Bettina Fazzinga, Sergio Flesca, and Francesco Parisi. Efficiently estimating the probability of extensions in abstract argumentation. In *Proc. of International Conference on Scalable Uncertainty Management (SUM)*, pages 106–119, 2013.

[22] Bettina Fazzinga, Sergio Flesca, and Francesco Parisi. On the complexity of probabilistic

abstract argumentation. In *Proc. of International Joint Conference on Artificial Intelligence (IJCAI)*, 2013.

[23] Bettina Fazzinga, Sergio Flesca, and Francesco Parisi. On the complexity of probabilistic abstract argumentation frameworks. *ACM Transactions on Computational Logic*, 16(3):22, 2015.

[24] Bettina Fazzinga, Sergio Flesca, and Francesco Parisi. On efficiently estimating the probability of extensions in abstract argumentation frameworks. *Int. J. Approx. Reasoning*, 69:106–132, 2016.

[25] Bettina Fazzinga, Sergio Flesca, Francesco Parisi, and Adriana Pietramala. PARTY: A mobile system for efficiently assessing the probability of extensions in a debate. In *Proc. of International Conference on Database and Expert Systems Applications (DEXA)*, pages 220–235, 2015.

[26] Anthony Hunter. Some foundations for probabilistic abstract argumentation. In *Proc. of International Conference on Computational Models of Argument (COMMA)*, pages 117–128, 2012.

[27] Anthony Hunter. A probabilistic approach to modelling uncertain logical arguments. *International Journal of Approximate Reasoning*, 54(1):47–81, 2013.

[28] Anthony Hunter. Probabilistic qualification of attack in abstract argumentation. *International Journal of Approximate Reasoning*, 55(2):607–638, 2014.

[29] Hengfei Li. *Probabilistic Argumentation*. PhD thesis, University of Aberdeen, King's College, Aberdeen, Scotland, 2015.

[30] Hengfei Li, Nir Oren, and Timothy J. Norman. Probabilistic argumentation frameworks. In *Proc. of International Workshop on Theorie and Applications of Formal Argumentation (TAFA)*, pages 1–16, 2011.

[31] Hengfei Li, Nir Oren, and Timothy J. Norman. Relaxing independence assumptions in probabilistic argumentation. In *Proc. of International Workshop on Argumentation in Multi-Agent Systems (ArgMAS)*, 2013.

[32] Diego C. Martínez, Alejandro Javier García, and Guillermo Ricardo Simari. An abstract argumentation framework with varied-strength attacks. In *Proc. of International Conference on Principles of Knowledge Representation and Reasoning (KR)*, pages 135–144, 2008.

[33] Sanjay Modgil. Reasoning about preferences in argumentation frameworks. *Artificial Intelligence*, 173(9-10):901–934, 2009.

[34] Juan Carlos Nieves and Roberto Confalonieri. A possibilistic argumentation decision making framework with default reasoning. *Fundamenta Informaticae,*, 113(1):41–61, 2011.

[35] Nir Oren and Timothy J. Norman. Semantics for evidence-based argumentation. In *Proc. of International Conference on Computational Models of Argument (COMMA)*, pages 276–284, 2008.

[36] Mauricio Osorio and Juan Carlos Nieves. Possibilistic well-founded semantics. In *Proc. of Mexican International Conference on Artificial Intelligence (MICAI)*, pages 15–26, 2009.

[37] David Poole. The independent choice logic for modelling multiple agents under uncertainty. *Artificial Intelligence*, 94(1-2):7–56, 1997.

[38] Henry Prakken. An abstract framework for argumentation with structured arguments. *Argument & Computation*, 1(2):93–124, 2010.

[39] Iyad Rahwan and Guillermo R. Simari, editors. *Argumentation in Artificial Intelligence*. Springer, 2009.

[40] Tjitze Rienstra. Towards a probabilistic dung-style argumentation system. In *Proc. of International Conference on Agreement Technologies (AT)*, pages 138–152, 2012.

[41] Matthias Thimm. A probabilistic semantics for abstract argumentation. In *Proc. of European Conference on Artificial Intelligence (ECAI)*, pages 750–755, 2012.

[42] Leslie G. Valiant. The complexity of computing the permanent. *Theoretical Computer Science*, 8:189–201, 1979.

[43] Bart Verheij. Two approaches to dialectical argumentation: Admissible sets and argumentation stages. In *Proc. of the International Conference on Formal and Applied Practical Reasoning (FAPR)*, pages 357–368, 1996.

Against Narrow Optimization and Short Horizons: An Argument-based, Path Planning, and Variable Multiattribute Model for Decision and Risk

Ronald P. Loui
University of Illinois Springfield, USA
`r.p.loui@gmail.com`

Abstract

This paper proposes a mathematical approach to analysis of decision and risk that makes use of the constructive argument logics that have become commonplace recently in artificial intelligence.

Instead of requiring an idealized, expected utility analysis of alternatives, in this paper, arguments appraise the desirability, comprehensiveness, and acceptability of incompletely described projections of the future. Instead of a qualitative risk management assessment process, threats and mitigations are represented numerically, but appraised with arguments, especially probability arguments and mitigation arguments, not averages. Arguments are given for or against the adequacy of commitments. Instead of using logic to derive the properties of acts that transform situations, *e.g.*, to construct goal-satisfying plans, in this paper, dialectical burdens are placed on demonstrating to a standard that investments and response policies will attain each milestone on a proposed trajectory. Trajectories are extensible and valuations are multi-attribute with varying completeness as knowledge permits. Superior trajectory specificity will be related to superior argument specificity.

The resulting picture of decision is a mixture of search, probability, valuation, and risk management; it should superficially bear a resemblance to satisficing mixed-integer discrete time control and many recent approaches to practical reasoning through argumentation. It is intended as an alternative to narrow optimization, which permits easy sacrifice of externalities on the grounds that they are hard to measure as real values. It is also intended as an alternative to fixed horizon decision-making, which produces unsustainable extremizations.

Keywords: Argument, Decision, Risk, Planning, Risk Management, Risk Analysis, Qualitative Decision Theory, Practical Reasoning, Sustainability, Artificial Intelligence, Satisficing, Standards

1 Introduction

For the past few decades, a separation between two kinds of analysis of choice has become entrenched. Everywhere, there are the disciples of decision analysis, chiefly emanating from microeconomic and statistical disciplines. They have lotteries, expected utility maximization, and preference axioms, and often call themselves Bayesian decision theorists. Elsewhere, there are the practitioners of risk analysis, mainly found in technology management and policy practices. They have threat assessment, hazard severity tolerance, and event chain methods, and are often associated with safety and reliability engineering, security, and regulation.

The entrenchment of these two fields, and the entrenchment of their separation, has led to some anomalies. Some attacks on expected utility are famous: some kinds of value are difficult to encode in an expected-utility model. But the anomalies this paper is concerned with are perhaps more pernicious. These anomalies have more to do with long-term societal distortion of values than one-shot personal contortion of preferences to axioms.

First, there is the anomaly of narrow optimization: a difficulty measuring and comparing value in multiple dimensions often results in important criteria simply being excluded from the model. As a result, some kinds of costs and returns are given precise mathematical expression, and these few are optimized at the expense of the many, less conveniently quantified, but equally important properties that have been excluded. Simply contrast 0-to-60 time, gigabytes per month, watts per channel, megapixels, renminbi per capita, body mass index, per cent minority, and aviation seating capacity, against water cleanliness, employee health care, alignment with mission, institutional and personal values, dignity, self-respect, and social justice. The latter will almost always disappear from the decision model, with no pressure to provide even penumbral accounting. A few trees will be optimized at the expense of the forest.

Second, there is an anomalous focus on attainment rather than sustainment: a difficulty determining commitments and valuations past a convenient horizon. This results in outcome descriptions fixed at a fictitious final time, so that long-term future disadvantage often pays for short-term advantage. Compare the median income after 10 years with life-long earnings; compare angioplasty to coronary artery bypass graft; or compare the overuse of a baseball team's bullpen staff to win one game versus the forward-looking management of reserves.

The decision theories of Savage [61] and Jeffrey [30] do not actually prevent modeling of value with long term horizons and broad perspectives on value. The objects of value are abstract. In fact, it is Raiffa who popularizes multiple objective utility models, with Keeney [33], and Fishburn follows suit [21, 22]. But the inconvenience of multiattribute models, due to nonlinear exchangeability, or worse, non-comparable dimensions and non-exchangeability, has resulted in narrow optimizations. This especially happens to discrete-valued attributes, in so-called mixed-integer models. This author believes that the unwillingness to model value in numerous dimensions is the result of these theories demanding too much precision: incompleteness of information and lack of precision are the norm for most attributes (see qualitative decision approaches to this same issue, *e.g.*, [70] and [16]). Ethical and legal aspects of decision, broad societal and environmental impacts, and distant future impacts, are especially difficult to characterize precisely. Similarly, the abstract concept of an outcome does not prevent an aggregate view over future possibilities, or even prevent infinite future envisionment. But it strongly dissuades too much consideration of the future: modeling precision correlates with short horizons. There must be more room in the model for real persuasion based on hard-to-characterize, imprecise aspects of decision.

Many admirable quantitative process models and semi-quantitative risk assessment approaches have been suggested in the huge literature of risk analysis. This includes the risk matrix (see [4]), goal structuring notation (see [34]), risk management plan [12], and even the mean-maximization subject to variance-limit portfolio methods [39]. Hundreds of articles appear in project management with similar sensibilities. This paper is sympathetic to the motivations of these approaches, though not necessarily to the logic or mathematics of their expression. These ideas have not found their way into many decision analyses because they have lacked an adequate symbolic representation. One milieux prefers process, while the other prefers precision.

2 Argument and Deduction

The development in logic that provides a new approach is the appearance of mathematical frameworks for pro-con, dialectical, defeasible argument, with the formal detail required for computation and artificial intelligence (AI).

These logics are non-deductive in the sense that demonstration is provisional: an argument does not prove its conclusion; justification depends on what counterarguments there might be. These logics determine conclusions, dependent on the state of argument production. Over time, as search for arguments proceeds, conclusions may be drawn, or not. They tend to be based on reasons that are defeasible, such as policies, or rules of thumb, that yield exceptions. In the current setting, the arguments are as likely to be analogical arguments from precedents, which have been of great importance in AI and Law (*e.g.* [5], [54]), or statistical arguments based on imperfect reference classes ([38] and [51]). No single mathematics of argument has yet emerged as canonical; the approach of P. M. Dung is popular [17] through its connection to logic programming; this author still uses his own framework [64] and [45] partly based on J. Pollock [50] and N. Rescher [60] (see surveys of mathematical defeasible argument [14], [56], and [10]; see also [28], [69], [53]).

Formally, a dialectical state for a proposition, p, is a set of arguments, $ARGS$, constructed by subsumption under a set of defeasible reasons $RULES$, or analogies to a set of precedent cases, $CASES$, with an evidential basis EV (facts or conditions not subject to dispute). The use of $ARGS$ is rule-based reasoning, while the use of $CASES$ is analogical reasoning, and the two can be mixed in a particular dispute. $ARGS$ contains at least one argument that derives p from EV using $RULES$ and/or $CASES$. An argument in $ARGS$ may have intermediate conclusions; if q is such an intermediate conclusion, then $ARGS$ may contain counterarguments to the arguments for q, that is, arguments for not-q. In some frameworks, the set $ARGS$ may also contain arguments that "undercut" application of a rule in $RULES$ (by exception) or (ir-)relevance of a case in $CASES$ (by distinction), or may contain priority meta-arguments that claim one argument defeats/dominates another. In our framework (and many others, see [18] and [19]), defeat is based on a syntactic criterion such as some kind of evidential or rule-based specificity (*i.e.*, some analogies are more specific than others; hence one defeats/dominates the other when two arguments are in opposition). Usually there is a way of calculating which of the arguments in $ARGS$ retain their ability to justify, which retain their ability to block justification, and which are simply defeated/dominated (and can be removed from $ARGS$ without changing what propositions are defeasibly justified). These calculations bear a striking resemblance to the original TMS of [15].

A probability argument might start with a reference class for an event, and a frequency of a property observed among that class. So 5 of 35 F's may have had h. Defeasibly, 5/35, or the interval constructed around 5/35 at some confidence level is $prob(h|F)$. A counterargument might be that the event in question is not only an F, but also a G. 14 of 18 G's have exhibited h. Since an interval constructed around 14/18 will disagree with an interval around 5/35 (*i.e.*, since neither interval will contain the other), all that can be said about the probability is the minimal containing interval, unless there is some method of combination. Of course, if 7 h's have been seen among 11 events that are F's and G's, the disagreement between argument based on F and argument based on G is replaced with the better argument based on the conjunction. It is important for the reader to see that probability arguments based on different, or subsuming, conditionalization can occur in a process of pro- and con- argument production; in a Bayesian setting, one conditions on the totality of evidence, but there may be dispute over the conditions themselves.

A precedent argument might start with a prior case that shares some properties with the current case, F and G. In that prior case, perhaps *not-h* was decided, such as the insufficiency with respect to a fiduciary standard, ethical standard, or due diligence. Perhaps the prior case weighed an argument based on F (pro-h) against an argument based on G (con-h). Absent a more specific precedent, in a new case where F, and G hold (equal specificity), or a new case where F, G, and E hold (increased specificity), one can argue *not-h* on this precedent. Moreover, even in a new case where just G holds, one can still argue *not-h*, by co-opting the argument from the prior case. There are more interesting analogies that pertain to the relevance of properties used in rebuttals and subargument rebuttals. If E was needed in the prior case to rebut the F-based argument, then E must be present in the current case, if F is present, in order to draw the same conclusion. [44, 43] Also, if not-E were decided in the prior case, the disanalogy based on E could support counterargument.

This author has used an argument framework to address the heuristic valuation of outcomes in decision trees [41, 40, 42], in particular to address the Savage small worlds problem [62]. The idea was that the outcome in a decision tree, "play-tennis-tomorrow" or "marry-the-girl" or "poach-an-egg" is often grossly underspecified: both underspecified with respect to detail and underspecified with respect to future commitments. Rather than insist on a single decision tree that considers various uncertainties, defeasibility permits one tree to improve another; it permits one heuristic evaluation to replace another. Provisional valuation of nodes was based on the properties that can be asserted with some certainty; further analysis of chance, deeper search, and more specifically described outcomes would improve provisional values (just as heuristic valuation permits deeper reasoning about a chess position in classic AI; see also [13]). The use here is a more ambitious embedding of choice in dialectical argument, formalizing elements drawn from mixed-valued multiattribute utility, AI planning, and risk mitigation.

The rest of this paper assumes that these argument frameworks will continue to mature and become familiar to a wide audience. It is actually possible that argument could be replaced by black-box machine learning, but that would be a very different philosophical approach, less normative perhaps.

3 Fox and Krause, McBurney and Parsons

Two related lines of thought in AI had the early insight to embed risk analysis in a logic of defeasible argument. John Fox and Paul Krause, in a series of papers with co-authors relate argumentation to medical decision and risk, especially cancer risks. [35], [36], [37], [24], [26], [23] This line of work tends to use qualitative assessments of likelihood, and anticipates further mathematical development of defeasible argument.

> "[T]here are situations in which a numerical assessment of uncertainty is unrealistic, yet ... some structured presentation of the case that supports a particular assessment may still be required, and is achievable. [C]haracterising risk [with] patterns of argument ... is not just an academic exercise in the use of argument structures. A solution to the problems outlined ... is of vital importance." (p. 393, [36])

[49] builds on these works, applying the Dung argument framework to the problem of defeasible outcome entailment, which is crucial to the present model. However their work formalizes argument semantics whereas this paper has more interest in characterizing value. [68] shares an interest in multi-attribute preference and the fundamental problem of incomplete knowledge (not precise probabilistic uncertainty, but genuine knowledge gaps). Their work takes an axiomatic approach to defeasibility and risk that is more complicated than what is envisioned here.

Meanwhile, Peter McBurney and Simon Parsons, in a second series of papers [46, 47, 48] presage much of the motivation for the formalism proposed here. They note the difficulties of doing quantitative risk analysis, and the prospects of argument in risk analysis:

> "[E]stimating and agreeing quantitative probabilities and utilities ... is not straightforward in most real-world domains. We are therefore motivated to explore qualitative approaches to practical reasoning, and this paper presents an application of argumentation to this end." (p. 2, [46])

> "Given the interactive nature of stakeholder involvement in risk assessment and regulation decision-making, we would anticipate that any adequate model of decision-making would draw on argumentation theory." (p. 23, [48])

These authors made clear that the qualitative arguments arising in risk analysis could be expressed in the new mathematical models of argument.

4 Amgoud and Prade, Atkinson and Bench-Capon

Over many years these lines were further developed as argument formalisms were applied with greater formal commitment by Amgoud and co-authors ([2], [58], [3], [1]) and simultaneously by Atkinson and co-authors ([6], [9], [65], [7], [11]). Their work researches argumentation and practical decision making in a way that incorporates many risk analytic themes.

[2] starts as "a first tentative work that investigates the interest and the questions raised by the introduction of argumentation capabilities in multiple criteria decision-making ... [that] remains qualitative in nature." [58] considers arguments about beliefs (entailments), arguments about desires (valuation), and arguments about partial plans (mitigations, horizons). [1] later refers to these as epistemic arguments, instrumental arguments, and explanatory arguments, noting that they can be mixed during deliberation.

[9] starts with a practical medical problem and uses a version of Dung's argument framework to argue "sufficient conditions," "alternative ways to achieve a goal G," and "different policies and perspectives." "Critical questions" lead to defeat of arguments, or to revision of plans. [8] inventories sixteen different kinds of attacks that could be represented formally in an argument formalism for a decision. Later work integrates desirable and undesirable effects directly into the representation of proposed actions [65] and an alternating transition system (which governs dialectic, but could also be seen as mitigations of risk).

[52] considers similar aspects of risk analytic argument, but highlights the dialogical aspects that are more procedural than what is considered here. Many other authors have put practical reasoning in a formal argument framework, but have not been as explicit about risk analysis. [66] is an example of a paper where the certainty of the value of a subsequent state is addressed within a system of argument for decision.

Although this paper is motivated independently from consideration of the fundamentals of risk analysis and decision, it can also be viewed as an extension of the works of these authors. These later authors reinforced and refined the claims that formal argument systems could bring practical reasoning, and some kinds of risk mitigation, into a new logic and representation.

From the earliest work, *e.g.*, [31] to the latest, most of the notational focus has been on the formal argument frameworks in support of practical reasoning, plan revision, and risk mitigation, in a way that would be appreciated by at least qualitative decision theorists, and in a style that would make precise the early intuitions of philosophers like [25]. It seems this point has been made. Our departure is to find a representation of the *objects* of argument, not the arguments themselves, in such a way that multiattribute utility decision theorists and risk analysts can see a bridge to their traditions and representations. Thus, we differ from these prior works in our assumption that the argument notation is of less interest than the representations of value, state, connectivity of state, and probability, about which one might argue.

5 Mathematical Representation

We start by defining the value space in which trajectories will be represented:

> ***Attainment space*** (or achievement space, or aspiration space), *A-space*, is an extensible d-dimensional product space of real-, real-interval-, discrete- and binary-valued dimensions.

Each dimension is named, $name_i(A)$.

> For example, for $d=3$,
> $name_1(A)$ might be real-valued *salary*,
> $name_2(A)$ might be discrete *rank*,
> and $name_3(A)$ might be binary *pre-tenure*.

It would be a mistake to think that the dimension d is fixed throughout an analysis or dialectic; the extension to other concerns as a result of inquiry is an important dynamic. However, at any point in argument, it is useful to consider the dimensionality of *A-space* to be fixed (but values unknown). Values in a dimension are often measurable, mostly in the reals, but they may be discrete, and also carry uncertainty such as *90%-chance-at-profit*, *likely-liable* or *arguably-unethical*. Linear order of values is presumed, as well as whether desirability is positive monotonic or negative monotonic.

> $m@t$ is a ***milestone at a time*** t in *A-space* only if m is a partly-specified vector v in *A-space*, where c components have values (the others are unspecified), $0 < c \leq d$. $dim(m@t) = c$, the count of components of the milestone that have specified values.

Milestones are an analogue of utility thresholds, or satisficing levels, or goals in goal programming.

> For example, a milestone might be:
> <$100k, ?, ?, Full-Professor>@2018.

Since dimensions have names, we can remove unspecified values and represent milestones in projected subspace with remapped names, *e.g.*, simply

> <$100k, Full-Professor>@2018

as a useful shorthand.

> v is a ***momentum*** in *A-space* only if v is a rate of change of prospect of a milestone w.r.t. time.

For real-valued dimensions, rate of change is a derivative w.r.t. time; for interval-valued dimensions, rate of change is an interval of derivatives w.r.t. time; for qualitative dimensions, rate of change is a qualitative value. Momentum is most closely allied with the ideas of persistence and inertia in AI planning (a species of default reasoning).

> A momentum might be:
> <$1k/yr, Annual-Review>.

$p(m@t)$ is a **prospect of a milestone** $m@t$ only if p is a $dim(m@t)$-dimensional measure of the satisfiability of $m@t$.

Each component of $p(m@t)$ is either a probability measure, an interval probability measure, a qualitative valuation of likelihood, or a characterization of the strength of supporting argument for attainment. Prospects are the analogue of subjective probability in DMUU and the risk matrix in risk analysis, but they also import the idea of defeasible and probabilistic planning from AI.

> A prospect of <$100k, Full-Professor>@2018 might be:
> <0.625, logical-possibility>.

I is a set of **investments** or **non-contingent commitments** only if I is a set of scheduled actions, $a@t$ to be taken at specific times.

Actions are irreducible named entities. Investments are the closest analogue to a plan in AI or decision in DMUU.

> A set of investments might be:
> { submit-to-journal@2015, travel-to-conference@2016 }.

Sometimes the investment is measured in the same units as attainment in a dimension, such as dollar cost; in such cases, the main difference between investments at a time (debits) and negative attainments (credits) is that investments are not tracked through time, while attainments are (investments occur at specific times, while attainments are charted at each time as part of the path); in this way, the model is more "bang-bang" control than continuous control.

CR is a set of **response policies** or **contingent commitments** only if CR is a set of <*event, I*> pairs where occurrence of an event triggers the paired (committed) investments.

Events may be named, but also may be un-named deviations from milestones or prospects of milestones. Events are irreducible entities. Contingent responses are closely allied with risk response planning and reactive/dynamic planning in AI. *Events(CR)* is the set of events considered in *CR*.

> A set of response policies might be:
> { <*rejected-article, resubmit-immediately*>,
> <*slashed-university-budget, add-external-consulting*>,
> <*negative-teaching-reviews, solicit-student-reference-letters*>
> }.

T is a **trajectory** only if $T = <M, V, P, I, CR>$ consisting of

1. a sequence of milestones $<m_0, ..., m_n>$, each $m_i = m@t_i$ (monotonic in times t_i),
2. a sequence of momenta $<v_0, ..., v_n>$,
3. a sequence of prospects, $<p(m_0), ..., p(m_n)>$,
4. a sequence of investments, $<I_0, ..., I_n>$, and
5. a set of contingent response policies, *CR*.

Note that the dimensionality of each milestone (or momentum or prospect) can change (usually diminish) as the sequence is extended. *T* has **length** *n* and **temporal extent** t_n.

> A set of **situational entailment arguments for (pro) a trajectory** *T*, is a set of arguments *pro-E*, where each subset *pro-E_i*, contains arguments for $p(m_{i+1})$ based on $p(m_i)$ and v_i.

Each argument, ARG_{ij} in *pro-E_i*, refers to a specific dimension shared by m_i and m_{i+1}.

If the argument refers to a reference class of instances that attained the subsequent target value (or range), it is a **frequency argument** (familiar to statistical inference; see [51]). If the argument refers to the causal efficacy of an investment, it is a **causal argument** (usually a defeasible planning argument familiar to AI planning; see [20]). If the argument refers to the adequacy of a contingent response, it is a **mitigation argument** (usually based on precedent and standard of proof or standard of demonstration, familiar to case-based reasoning in AI and law; see [27], [55]).

A set of *situational entailment arguments against (con) a trajectory* T, is
a set of arguments $con\text{-}E$, with subsets, $con\text{-}E_{ij}$, where each member attacks
some argument in $pro\text{-}E_i$.

An **attainment dialectic over a trajectory** is a set of situational entailment arguments pro and con together with an appraisal of the dialectical state, *i.e.*, describing the strength of the arguments for the trajectory's plausibility meeting a standard of proof/demonstration. This description includes the current judgment pro or con, any currently persuasive arguments, and the list of effectively rebutted and unrebutted counterarguments.

Optimality is replaced with the concept of

1. better argument
   (including dialectical strength over more comprehensive *events(CR)*; more mitigation equals better argument),

2. more robustly characterized milestones (greater c),

3. higher attainment returned on acceptable investment, and

4. more extended sustainment (greater n).

Clearly this is a constructive approach rather than an exhaustive search of the space of choices. The desire to have robust envisionment in *events(CR)* is the crux of risk analysis. The desire to relate attainment to investment is the basic instinct of utility. The desire to enlarge c is the basis of multicriterion analysis. The extendable trajectory emphasizes sustainability.

The initial construction problem is thus:

A trajectory T is **justifiable to standards** $\langle p^\dagger(t), I^\dagger, CR^\dagger \rangle$ w.r.t. a set of arguments $E\text{-}pro$ (surviving adversarial finding of $E\text{-}con$) only if the attainment dialectic justifies the trajectory under these standards:

1. all $p_o(m@t) \geq p^\dagger(t)$, (the prospective standards, which may vary as a function of time)

2. all $I_i \leq I^\dagger$, (the investment standards) and

3. all $\langle e, r \rangle$ pairs in CR meeting the $CR^\dagger$ standard according to precedent (the mitigation standards).

Standards for prospect are allowed to decrease as distant futures are considered because it is difficult to establish much of a guarantee. The alternative, which is not necessarily a bad alternative, is to let future milestones be probabilities of attainment, so that prospects are meta-probabilities, or meta-arguments, that can meet higher standards.

These standards, together with the membership of CR, the levels of attainment at each milestone, the specificity of each milestone (more attributes are better), and the length of the trajectory, are the known **performance measures** of T.

The improvement problem is iterative:

> Given a trajectory T^0 justifiable to standards $\langle p^0(t), I^0, CR^0 \rangle$ with milestones $< m_0, ..., m_{n0} >$, find an alternate or extended trajectory T^1 justifiable to those same standards with superior performance measures, that is, for each $m_i @ t_i$ in T^0 there is a corresponding $m_j @ t_j$ in T^1 such that $m_i \geq m_j$ where $t_i = t_j$.

Note that it may be easy to find alternatives to T^0 that are arguable to non-comparable standards or non-orderable performance measures, but improvement requires superiority: equivalence at least in all dimensions, and superiority in at least one. Also note that if T^1 is justifiable to the standards of T^0, it may yet be justifiable to higher standards; but a ratcheting to higher standards of justification may not be desirable in the search to improve performance measures.

As is typical in constructivist settings, search can be directed in multiple places: either at improving *E-pro*, improving *E-con*, or finding an improvement T^1 to an existing T^0. One can consider all three searches being performed in parallel, or according to a specific protocol (including possibly randomization or some kind of search optimality).

Arguments and precedents require additional representation. In this paper, the claims and counterclaims of argument are given, but not their formal representation. A formal system of argument presupposes many rules (a rule base), and arguments from precedent require many cases (a case base). Many details can be found in the references already provided. Here the focus is on notation for the objects of argument, not the arguments themselves.

6 How Can This Diverse Syntax Be An Improvement?

The improvement is in the representation: a trajectory of measurable milestones with arguable valuation, arguable entailment, arguable length, detail, and dimensionality, orderable but not necessarily scalable measurements of value, and arguable standard of hazard mitigation. It puts a spotlight on concerns that have been missing from the entrenched models. It puts front and center many of the concerns that are important, but get omitted because they are hard to include. It should be constantly clear that the trajectory is not a final or exhaustive depiction of the situation. It is the scorecard of an argument game aimed at uncovering hidden criteria and constraints, forcing commitments and contingency planning, and revealing downstream consequences. It presupposes an adversarial process of search for improved appraisal: search can lead to more detail, lowering of aspiration, and discussion of hazards. It is very much what project management planning risk analysts do during semi-quantitative risk analysis, except that here, one can write the specific arguments from reference classes and precedents, and usually determine what kind of response would suffice. It should lead risk analysts to more numerical specificity, more plan flexibility, more transparent standards of justification, and higher levels of aspiration.

The improvement over multicriterion decision analysis is mainly the step away from a fiction of neatness, completeness, and precision. Anyone can draw a bad decision tree, but even good decision trees can leave difficult questions unanswered. Here, in the enumeration of multiple criteria, one need not produce weights, or a nonlinear mapping, for combination and dimensionality reduction. In the enumeration of chance events, one need not be exhaustive, and one need not attach a probability to each event. Decision trees grow large when chance, choice, and time are all increasing the node count, and nodes beget more nodes.

As pointed out in the author's earlier work, there is no place in the classic theory for refinement of a decision tree through search; this is because the axioms of preference over lotteries require that all nodes have a value that already summarizes the expected value of any subtree extending from that node. This has been a lot to ask of the decision analyst: every subtree must be perfect with respect to further elaboration. Embedding the analysis of value (heuristic valuation of utility, like chess configurations) in argument was the author's earlier move, and that work suggested the potential for arguing probability values as well.

This work takes those arguments over decision trees and replaces expected heuristic utility with claims of time-indexed attainments, and the burden of arguing entailments that connect milestones. It explicitly makes room for the dialectical interplay that underlies risk analysis processes.

Probability has been removed from a discounting of value to a connection of milestones. This is like asking that aspiration meet a probabilistic standard of confidence or acceptability. It is possible to represent upside and downside risk discounted by probability, but it is not as easy to represent as the probability argument of attainability of one milestone from the previous milestone.

6.1 Representing the Classic Lottery Example as a Simple, Degenerate Case

First, consider how a classic lottery might be represented in this notation.

A ticket can be purchased for $1 that has a 1/10 probability of paying $20. Of course, this model requires that one be specific about the time of entry and time of payoff in order to represent milestones.

The naïve representation shows only the milestone representing probable loss, on a single dimension of dollar attainment:

$T^0 = <$
  \# milestones $< <\$0>$ @lottery-payout-time $>$,
  \# no momenta $< >$,
  \# prospects $< <.9> >$,
  \# investments $\{ \$1$ @lottery-entry-time $\}$,
  \# no events-responses $\{ \}$
$>$

But a refinement increases the dimensionality of milestones, looking at both gain and loss:

$T^1 = <$
  \# milestones $< <\$0, \$20>$ @lottery-payout-time $>$,
  \# no momenta $< >$,
  \# prospects $< <.9, .1> >$,
  \# investments $\{ \$1$ @lottery-entry-time $\}$,
  \# no events-responses $\{ \}$
$>$

and the relative desirability of this trajectory compared to the alternative of not entering the lottery at all depends on arguments that presumably take into account risk aversion, time value of money, and opportunity costs. That is, if the probability-discounting of gain and loss is to be done without transformation into utility, and the two combined to produce a dollar-expectation, that is something that is an argument for relative desirability, or an argument for meeting a standard of attainment, not an automatic reduction of dimensionality, time, and the serendipity of numerical prospect. For all we know, the next argument move might be to argue that the value of 0.1 prospect of attaining $20 is incorrect. One counterargument is thus non-entry into the lottery,

$T^2 = \,<$
   # milestones $<$ $<\$0, \$0>$@lottery-payout-time $>$,
   # no momenta $<\,>$,
   # prospects $<\,<1,\,1>\,>$,
   # investments $\{$ $\$0$@lottery-entry-time $\}$,
   # no events-responses $\{\,\}$
$>$

and a refinement of the argument to enter the lottery might be the extension of attainment to include expected gain:

$T^3 = \,<$
   # milestones $<$ $<\$0, \$20,\,expected\$2>$@lottery-payout-time $>$,
   # no momenta $<\,>$,
   # prospects $<\,<.9,\,.1,\,1>\,>$,
   # investments $\{$ $\$1$@lottery-entry-time $\}$,
   # no events-responses $\{\,\}$
$>$

Here, one might argue from precedent that an expected-$2 gain on $1 investment returned over the difference in time meets the standard of fiduciary responsibility, portfolio aspiration, prudence, or best-known-use of cash on hand, etc. Note that a counterargument might consider the variance, not just the mean, or the nonlinearity of dollar utility, etc.

6.2 Representing the Classic Risk Example as a Simple, Degenerate Case

Next, consider how a classic risk analysis mitigation might be represented in this notation.

A nuclear power plant can be built in 1985 to supply power for 50 years, but there is a small probability of disaster.

$T^0 = \,<$
   # milestones $<$ $<$sufficient-power-for-past-50-years, huge-disaster$>$@2035 $>$,
   # no momenta $<\,>$,
   # prospects $<\,<.99,\,.000001>\,>$,
   # investments $\{\,\}$,
   # no events-responses $\{\,\}$
$>$

One might argue directly about the acceptability of reaching a milestone at 2035 with a 10E-6 probability of a "huge disaster," which may actually depend on quantifying the disaster. But an obvious refinement is to commit to a mid-term inspection, which reduces the probability (presumably a probability argument is given for the new conditionalization) to 10E-7:

$T^1 = <$
    # milestones $< <$sufficient-power-for-past-50-years, huge-disaster$>@2035 >$,
    # no momenta $< >$,
    # prospects $< <.99, .0000001> >$,
    # investments { inspection-and-repair@2010 },
    # no events-responses { }
$>$

and 10E-7 might meet the standard of acceptable risk. Even better would be to commit to adoption of any new technology safety upgrades (which may or may not change the probabilities):

$T^1 = <$
    # milestones $< <$sufficient-power-for-past-50-years, huge-disaster$>@2035 >$,
    # no momenta $< >$,
    # prospects $< <.99, .0000001> >$,
    # investments { inspection-and-repair@2010 },
    # no events-responses { $<$new-safety-technology, implement-it$>$ }
$>$

6.3 A Larger Used Car Example

Consider the purchase of a used 2002 Toyota Highlander purchased in 2015 for $8000 with 106,000 miles on the odometer. Here, we show argument moves, not just representation. The less-than-certain initial prospects reflect the nontrivial possibility that the car does not meet manufacturer's specifications. The first proposition is that it can be driven for 10 years with no further investments.

$T^0 = <$

    # milestones

    $<$

    <$8k or less cost, 18-mpg or more, 494g/mi or less CO_2, full-time-4wd, 13 years old, 38.5 cu ft, 106kmi or less odo>@2015,

    <$18k or less cost, 18-mpg or more, 494g/mi or less CO_2, full-time-4wd, 23 years old, 38.5 cu ft, 206kmi or less odo>@2025

    $>,$

    # momenta

    $<$

    < +$1k or less repair/yr, 0, 0, +1 years old/yr, 0, +10kmi or less odo/yr >,

    < +$2k or less repair/yr, 0, 0, +1 years old/yr, 0, +8kmi or less odo/yr >

    $>$

    # prospects

    $<$

    <1, .95, .95, .95, 1, 1, 1>,

    <.6, .85, .85, .95, 1, 1, .85>

    $>,$

    # no investments { },

    # no events-responses { }

$>$

There might be fairly good statistical arguments that such automobiles reach the 2025 cost, emissions, and mileage efficiency based on age and/or mileage. The mileage number itself is prospective, so its uncertainty might be propagated in some models. No uncertainty is attached to the momenta, though the justification of a subsequent milestone's prospects might depend on the justification of the momentum values.

That is, one way to counterargue the proposition that the trajectory is justified might be an argument against the momentum estimate. In fact, the strongest counterargument to attaining the 2025 milestone, even with the low .6 prospect standard, would be to suggest that the cost momentum grows, and could be reflected more precisely using an intermediate milestone.

An immediate rebuttal to such counterargument might be that a major service will be done in two years. This still changes the cost milestones, and does not require an intermediary, but may solidify the case for prospects of entailment.

Another response might be to lower the mileage milestone, adding an inflection point at which the miles per year are reduced (T^2).

$T^1 = <$
    # milestones
    $<$
        $<$\$8k or less cost, 18-mpg or more, 494g/mi or less CO2, full-time-4wd, 13 years old, 38.5 cu ft, 106kmi or less odo$>$@2015,
        $<$\$21k or less cost, 18-mpg or more, 494g/mi or less CO2, full-time-4wd, 23 years old, 38.5 cu ft, 206kmi or less odo$>$@2025
    $>$,
    # momenta
    $<$
        $< $ +\$1k or less repair cost/yr, 0, 0, +1 years old/yr, 0, +10kmi or less odo/yr $>$,
        $< $ +\$2k or less repair cost/yr, 0, 0, +1 years old/yr, 0, +8kmi or less odo/yr $>$
    $>$
    # prospects
    $<$
        $<$1, .95, .95, .95, 1, 1, 1$>$,
        $<$.8, .9, .9, .95, 1, 1, .85$>$
    $>$,
    # investments { major-\$3k-service@2017 },
    # no events-responses { }
$>$

$T^2 = <$
    # milestones
    $<$
        $<$\$8k or less cost, 18-mpg or more, 494g/mi or less CO2, full-time-4wd, 13 years old, 38.5 cu ft, 106kmi or less odo$>$@2020,
        $<$\$8k or less cost, 18-mpg or more, 494g/mi or less CO2, full-time-4wd, 13 years old, 38.5 cu ft, 106kmi or less odo$>$@2015,
        $<$\$21k or less cost, 18-mpg or more, 494g/mi or less CO2, full-time-4wd, 23 years old, 38.5 cu ft, 181kmi or less odo$>$@2025
    $>$,
    # momenta
    $<$
        $< $ +\$1k or less repair cost/yr, 0, 0, +1 years old/yr, 0, +10kmi or less odo/yr $>$,
        $< $ +\$1k or less repair cost/yr, 0, 0, +1 years old/yr, 0, +5kmi or less odo/yr $>$,
        $< $ +\$2k or less repair cost/yr, 0, 0, +1 years old/yr, 0, +3kmi or less odo/yr $>$
    $>$
    # prospects
    $<$
        $<$1, .95, .95, .95, 1, 1, 1$>$,
        $<$.9, .95, .95, .95, 1, 1, .95$>$,
        $<$.8, .9, .9, .95, 1, 1, .85$>$
    $>$,
    # investments { major-\$3k-service@2017, reduce-miles-per-year@2015 },
    # no events-responses { }
$>$

Of course, the cost of repair is not the only cost of ownership. One major hazard for car ownership is accidental collision. This is an opportunity to have a mitigation argument.

For the event *accidental-collision*, does the *$500-deductible-insurance* meet a reasonable standard? Perhaps it meets the standard required by state law, or the standard of prudence for a person of means, or a standard of *erring-on-the-side-of-caution*. There might be a simple table relating mitigations to standards for this common hazard. But there might be room for argument based on particular details, such as the fact that the auto is an *SUV-with-curb-weight-near-3500-lbs*.

T^3 reflects the inclusion of a preemptive commitment, where the cost of the insurance enters the milestone accounting and momenta, but the scenario of insurance coverage and deductible do not, because presumably the probability estimate would be too coarse to be meaningful at this stage. An actuary might produce a specific probability, say .001 over ten years, for a specific damage level, but the standard of mitigation might not be about calculable expected loss: legal requirement might be the dominant concern, or an ethic of taking responsibility comparable to the ethic belonging to others held in high esteem.

$T^3 = <$
 # milestones
 <
 <8k or less cost, 18-mpg or more, 494g/mi or less CO_2, full-time-4wd, 13 years old, 38.5 cu ft, 106kmi or less odo, $500-deductible-insurance>@2020,
 <18.5k or less cost, 18-mpg or more, 494g/mi or less CO_2, full-time-4wd, 13 years old, 38.5 cu ft, 106kmi or less odo, $500-deductible-insurance>@2015,
 <26k or less cost, 18-mpg or more, 494g/mi or less CO_2, full-time-4wd, 23 years old, 38.5 cu ft, 181kmi or less odo, $500-deductible-insurance>@2025
 >,
 # momenta
 <
 < +1.5k or less cost/yr, 0, 0, +1 years old/yr, 0, +10kmi or less odo/yr >,
 < +1.5k or less cost/yr, 0, 0, +1 years old/yr, 0, +5kmi or less odo/yr >,
 < +2.5k or less cost/yr, 0, 0, +1 years old/yr, 0, +3kmi or less odo/yr >
 >
 # prospects
 <
 <1, .95, .95, .95, 1, 1, 1>,
 <.9, .95, .95, .95, 1, 1, .95>,
 <.8, .9, .9, .95, 1, 1, .85>
 >,
 # investments { major-$3k-service@2017, reduce-miles-per-year@2015 },
 # events-responses { <major-accidental-collision, file-insurance-claim>}
 >

7 Additional Discussion of Elements I: Clarifications

Attainments are different from sustainment of attainments. In each dimension, an attainment is reported or specified in native or natural units. There is no need to transform levels of attainment into real-valued utility because there will not be combination with probability, nor combination across dimensions.

A-space dimensional values are used for orderability, not scalability (unless some argument permits a rate of exchange). Probability should enter as a standard for arguability of attainment. "That's not attainable, given the investments, the momentum, and the standard $p^\dagger = \langle 0.955, 0.95,$ *morally-certain, preponderance-of-evidence, from-on-point-case*$\rangle$." Expected utility may make sense when probability and utility can be measured with precision. But when probability itself can be argued, based on reference class disagreement, even a quantitative threshold such as 0.955 may reek of false precision. Discrete and binary valuation might make probability even less useful in many dimensions.

The dynamics are in the argument over meeting these standards. "On the contrary, it is attainable, at least to the standard $p^\dagger = \langle 0.85, 0.9,$ *as-certain-as-the-interest-rate-forecast, more-likely-than-not, from-relevant-case*$\rangle$." If that standard seems too low, perhaps there is an alternative trajectory with better performance measures that can be justified at that same standard or higher.

It is true that permitting extensibility of *A-space* and unspecified values in dimensions makes a milestone like a set of properties that might be proved of a resultant state in an AI planning representation. But trajectories of milestones, which are fully specified in a reduced set of dimensions are more easily visualized. Logical representations of future states may be too expressive to give the impression of controlling motion in *A-space*. One of the distinctive aspects of this model is that action might produce momentum, not just state change. The analogy to kinematics, and physical control, while still permitting incomplete information, is fundamental. The difference from mixed-integer control theory is that argument and justification replace expectation and optimization.

An important role for argument is connecting the milestones, arguing that the investments of non-contingent actions suffice to achieve the subsequent milestone with the prospect as claimed. There is the suggestion of a Markov process here, because situational entailment arguments are dependent on prior state, momentum, and action, not on the path to the prior state. But this is just a suggestion, since arguments may take into account more than prior state.

More important is the idea that *situation + action* does not automatically equal a precise subsequent situation. In AI planning, this historically began as a deductive entailment: the subsequent state being named as action applied to prior state, and the outcome properties at that state being derived. This subsequently led AI planning research to probabilistic and defeasible approaches to description of the outcome. In decision trees with utilities on outcomes, this connection is perhaps taken too much for granted. If there were any uncertainty about the result of action, the outcome state could be bifurcated by a chance event. But this still suggests that it is easy to know the entailments of actions, and that downstream chance bifurcation does not actually cloud utility estimation. In this model, the problem of arguing that *situation + action*, or *milestone + investment*, yields a particular successor situation is central.

Perhaps the easiest argument of situational entailment is that left alone, the momentum and time will cover the difference. Note that this presumes persistence of momentum. An easier argument might be directly frequentist: "90% of retirement accounts worth $500,000 by age 50, with continued contribution of 3% and 3% match, achieved $1M by age 65." A counterargument might be that there is a more specific reference class of retirement accounts owned by people with young children at age 50, where the percentage that move to the next milestone is more like 50%.

A typical counterargument to entailment is derailment or deviation through hazard: "What if the employer stops making matching contributions?" "Then there is a commitment to change employers or add consulting work." Rather than penalize the prospect of attainment because of the contemplation of this event, a mitigation argument is given, and a contingent commitment (like a control policy) is added to the trajectory. Similarly, a counterargument could incur a non-contingent commitment: "What if equities have substandard performance for a decade?" "There is a non-contingent (preemptive) commitment to diversify." Again, rather than change the prospect of attainment, the counterargument causes the trajectory to be altered by adding an investment. "What if the government changes the rules for tax-free accrual?" Here, there is no response; this can be charged against the prospect of attainment, or can lead to a low standard of CR adequacy (next paragraph).

Note that in this model, hazardous events do not have to be highly improbable, disastrous events; they need only be potential deviations worthy of response.

In those cases where very damaging events with very low probabilities are considered, the model includes them in the membership of CR, which meets some $CR^{\dagger}$ standard that is one part of performance. The event-response pair <*earthquake-power-outage, generator-backup*> is not as good as <*earthquake-power-outage, four-channel-independent-generator-backup*>, though one must now argue the acceptability of the increased cost of better mitigation. These are argued from precedents, with standards such as *minimal-liability-coverage*, *best-effort*, *meets-regulations*, *meets-best-practices*, *meets-company-standard-for-ethical-behavior*, or *no-worse-than-US-nuclear-power-safety*. These arguments can be very direct from on-point precedents, or merely remote analogies. Obviously cost of mitigation is important. But what we cannot do is draw a decision tree and try to estimate a very small probability and a very large disutility, then take an expectation.

Since the standard for acceptability often resides in a domain of public or legal appraisal of ethics or liability, arguing what standard can be met appears to match the model to the reality. The question of having a backup plan, and allocating those resources, is sometimes more important than pricing that backup plan when hazards are out of scale. Having a backup plan for *earthquake-power-outage* and *tsunami-damage-to-generators* is even better (superiority being reflected in inclusion, not merely cardinality). Having lower-cost generators is not an immediate concern. In the chess analogy, we are ignoring some pawn moves.

Still, the milestones in a trajectory can and should reflect the costs of investments and resources needed to meet contingent commitments. For example, if one dimension of attainment space is *cash-on-hand*, and there is an investment commitment of $100k in real estate funds as part of a diversification action in 2015, then the $100k difference should be reflected in the transition from one milestone (2015) to the next (2016), lest the prospect of entailment depend on magic. Similarly, if CR contains <*short-positions-called-in, cover-with-$30k-liquid-assets*>, then *cash-on-hand* or *line-of-credit* might require an increase along the trajectory. The determination of the adequacy of a response is linked to the positions reflected in the milestones.

Note that one could find a precedent case where more than $30k was required to cover short positions, in which case the argument for the adequacy of the response meets a lower $CR^\dagger$ standard. Also, one could argue that the transition from one milestone to the next has poor prospect if all of the commitments in CR are to be maintained.

The important observation here is that the costs of commitments are reflected in the milestones, but a numerical accounting of degree of mitigation, disutility under disaster, and probability of worst cases is avoided; this critical appraisal is relocated to the dialectic of entailment and performance standard, recognizing that much of the information will be incomplete (whether from imperfect epistemics or vagueness of description).

A simpler numerical case that showcases the mathematical expressiveness of this approach might be something like counting beans that might also be eaten, with no use of momenta.

7.1 Bean Counting as a Simple Example of Notation

$T^0 = \,<$
  # milestones
  $<$
    $<$0 counted, 10 uncounted, ? eaten$>$@noon,
    $<$1 counted, 8 uncounted, 1 eaten$>$@dinner,
  $>$,
  # no momenta $<\,>$,
  # prospects $<\,<$1, 1, ?$>$, $<$.9, .9, likely$>\,>$,
  # investments { start-counting-at-3pm, eat-breakfast },
  # events-responses { $<$fall-asleep, set-alarm$>$ }
$>$

The first argument in *pro-E_1* is purely statistical: 90% of similar counting cases, starting at 3pm, result in a bean counted by *dinner*. A second argument is qualitatively probabilistic, perhaps subjective: around *dinner*, eating one bean is *likely*; eating one bean, given that breakfast was eaten, is *likely*. These two single-condition (marginal) probability arguments would be defeated by a single two-condition (joint) probability argument. The reference class might also be reduced to the subset of hedonistic bean counters, who are likely to eat more than one bean around *dinner*. Such an argument would enter *con-E_{12}* but we would like the trajectory to be justified at first.

As there is no cost or effort dimension, there is no place to reflect the mitigation actions. Nevertheless, $CR^\dagger$ might be a *highly-effective* standard of mitigation, and the prospect of having counted the bean by *dinner* might even be argued to exceed .9 under a Bayesian argument.

Rather than counterargue, perhaps T^0 is accepted at its standard of justification, but a new trajectory proposed as improvement:

$T^1 = \,<$
  # milestones
  $<$
    $<$0 counted, 10 uncounted, ? eaten$>$@noon,
    $<$2 counted, 7 uncounted, 1 eaten$>$@dinner,
    $<$3 counted, 6 uncounted, ? eaten$>$@midnight
  $>$,
  # no momenta $<\,>$,
  # prospects $<\,<$1, 1, ?$>$, $<$.9, .9, likely$>$, $<$.8, .8, ?$>\,>$,
  # investments { start-counting-at-1pm, eat-breakfast },
  # events-responses { $<$fall-asleep, set-alarm$>$, $<$lose-a-bean, buy-a-bean$>$ }
$>$

That is, by starting counting two hours earlier, one can argue the prospect of having two beans counted by dinner to the same level of probability. Moreover, the standard for risk mitigation is higher because a second hazard has been addressed. It was easy to construct an improved set of milestones because there is not yet an accounting of cost and effort. This trajectory also steps further into the future, so it has superiority in that respect.

7.2 Retirement Cash Drawdown Example

This example refers to the planning of funds used in retirement for assisted living, where the sale of a major asset needs to be timed well, there is an interval-valued forecast of return on real estate, and there is high burn rate for cash.

$T^0 = <$
 # milestones
 $<$
 $<$\$200k cash on hand, \$800k real estate valuation$>$@2014,
 $<$\$130k cash on hand, [\$840k, \$920k]$>$@2015,
 $<$\$60k cash on hand, [\$848k, \$924k]$>$@2016
 $>$,
 # momenta
 $<$
 $<$-\$70k/yr, [+5%, +15%]$>$,
 $<$\$70k/yr, [+1%, +10%]$>$,
 $<$-\$70k/yr, [-5%, +15%]$>$
 $>$,
 # prospects
 $<$
 $<$1, 1$>$,
 $<$likely, likely$>$,
 $<$likely, likely$>$
 $>$,
 # no investments { },
 # events-responses
 { $<$return-drops-below-3%, sell-real-estate-asset$>$, $<$cash-on-hand-below-\$30k, sell-real-estate-asset$>$}
$>$

The counterargument that might be helpful here is that the proposed asset sale as mitigation of low cash on hand is inadequate because of illiquidity. The argument might refer to the range of time it takes to sell property of this kind. A typical financial decision model might consider the average time to sell, with a forecast average return on real estate, and try to calculate the latest time to sell based on positive expected investment growth prior to the last-sell-date.

A better model would consider bounds at acceptable risk levels: not just the averages, but the time to sell at an acceptable probability, and the return on real estate lower bound at an acceptable probability. A risk analysis might sell too early, failing to recognize that some better mitigations might permit higher attainments. It tends to be defensive rather than generative. Here, attainments can be ratcheted higher. For example, having an offer in hand, even below market price, might permit *cash-on-hand-below-$30k* event, or having a pre-approved line of credit on the asset might even be better, permitting a longer holding of the asset.

While optimality is not a part of this model, improvement of attainment, and prospect of attainment, perhaps at lower cost, and for a longer time, is shared with the utility-maximizing crowd.

8 Additional Discussion of Elements II: Comparisons

Some will want to compare this model to goal-oriented AI planning. Just as in AI planning, much attention is given to the succession of states and the component-by-component, possibly incomplete description of successor states.

In AI, there is a logical derivation of properties, usually with an emphasis on the derivation from logical expression of precondition, action, and effect. This derivation may use default rules of causal efficacy, or persistence of properties under non-action. These are reflected in the possibility of causal entailment argument here. AI planning may include probabilities of properties, which is like a dual of probabilistic argument of succession, with an emphasis on the uncertainty of a *Holds(p, a-applied-to-s)* relation rather than the uncertainty of a *Leads-To(<p, q>, <p', q'>)* relation. AI robotics path planning also includes Markov Decision Processes based on state transition probabilities (see for example, [32] and [29]). The author does not know of AI planning models that consider hazard mitigation with the same importance it is given in risk analysis (though see [63] and other work using AI planning for security).

In this model and AI planning, the emphasis is on achieving specified goals by committing to action. The costs of those actions, and the relative desirability of various partial goal satisfaction, are not normally subjected to utility analysis. Thus, a plan that achieves three goals in five steps, using four actions, does not receive a score that can be compared with a plan that achieves two different goals in four steps, using eight actions.

For the same reason, it may be a mistake to think of investments as reductions of value. In AI planning, if an action is found that produces the desirable outcome, the relative cost of the set of actions is rarely quantified and subtracted from the value of its entailments. This could be a flaw in the set-up, or a failure to value the efforts of an automaton (but not all AI planning problems are intended to be executed by computers or robots). Or it could be a different perspective on the interplay of attainment and commitment, where the appraisal of commitments is based more on utilization, obligation, responsibility, or free exercise of will.

For those who routinely use utility to compare n apples to m oranges, the reluctance to rely on exchange rates, to price actions, time, and policy commitments, may seem a step backwards. How can one compare high achievement at one cost against higher achievement with an additional cost? How can one compare more apples and fewer oranges against fewer apples and more oranges? The short answer is that we do not attempt to. It may be the illusion of numeric exchange rates that has distorted choice theory in the first place. In an era where aspirations are personal and experiential, like backrubs instead of widgets, where liabilities are dyadic and contextual rather than universal, it is hard enough to describe the objects of value, much less measure their intensity on ratio scales, and worse, to find among the multitude of scales a projection into linear order.

For those who routinely discount utility by probability (or by time), the reluctance to use probability to reduce aspiration levels by multiplication may seem an irrational abstinence. Why not penalize the aspiration by the prospect of its attainment, in proper proportion? What good is a numerical probability except to produce an expected value?

The answer is that not all decision making takes place in casinos with exhaustively known possibilities, physically-based chance mechanisms, and the law of large numbers. Epistemic probability is a guide to belief, but might not be a good way to reduce the levels of attainment. Subjective probability may have been too good to be true, as "that kind of probability that permits betting on proportionally discounted values." Probability arguments, and unbounded unknown unknowns, lead to forbearance here. The future, or its perception, may be full of uncertainty, but that does not mean it is full of lotteries. Even Abraham does not look upon Ishmael and see a 50% lottery on Isaac.

Risk methods often take some metrics as constraints and optimize other attributes subject to the satisfaction of that constraint. Here, the constraints are adopted on the adequacy of argument, and search for improvement of what can be justified replaces the requirement of optimality.

What about upside risks? Risk analysis tends to look pessimistically at the downside risks, to counter uncriticized optimism. But sometimes there is an upside opportunity. If it can be expressed as a highly probable attainment, or even an arguable possibility under a weak standard, it could be put in a milestone. Many gains however are improbable. In those cases, there is a choice of putting a probability or expectation of a gain in *A-space*, as a binary-valued attainment, with very good probability. Then the evaluation of this attainment, as a part of performance, depends on ancillary argument: perhaps there is a table directly showing cost and expectation tradeoffs that are justifiable. Thus, the arithmetic of expected utility can sometimes be included and can be informative and persuasive.

Another choice is to represent the upside risk as a mitigable event, where the mitigation is actually a plan for what might be done with that gain. In cases where utility is linear in money, where there is no penalty or reward for process, and prices and rewards are both in dollars, the mediation of exchange through argument will seem cumbersome. But replace the reward with something less pecuniary, such as *a-uniquely-pleasant-backrub* or *a-personally-guided-trip-around-the-world*, and the use of argument to mediate the cost and chance reward makes perfect sense.

8.1 A Different Argument Over a Lottery Ticket Example

Consider again the decision to buy, or not, a lottery ticket, this time for \$1 with a 1/1000 prospect at \$995, tomorrow, to be drawn next week. This time, the accumulated dollar cost of the lottery is tracked as part of the milestone:

$T^0 = <$

\# *milestones*

$< <$-1 *dollar, chance-at-\$1000*$> $@*next-week* $>$,

\# *no momenta* $< >$,

\# *prospects* $< <1, 1> >$,

\# *investments* { *buy-ticket@tomorrow* },

\# *no events-responses* { }

$>$

This can immediately be refined by extending the length of the trajectory:

$T^1 = <$
\# *milestones*
$<$
$<$-1 dollar, chance-at-$1000> @next-week,
$<$-1 dollar, chance-at-$1000> @next-week-and-a-day
$>$,
\# *no momenta* $< >$,
\# *prospects* $< <1, 1>, <.999, 0> >$,
\# *investments* { buy-ticket@tomorrow },
\# *no events-responses* { }
$>$

One could argue that a week's worth of having a chance at $1000 can be traded for $1 from some precedent argument (that argument would appear to require an earlier milestone showing the *chance-at-$1000* starts tomorrow). Presumably there are also precedents, or rules, supporting the non-exchangeability of $5 or $10 for the one week enjoyment of possibility. This would reflect some of the actual reasoning of people who derive process value. Counterargument could also look at the prospect of the lottery being unfair, or a scam, or unlikely to be payable (the current situation with the Illinois state lottery during budget stalemate).

One could add the dimension of a specific *0.1%-chance-at-$1000*, or a specific *expected-99.5-cent-win*, the former representation permitting a risk-affinity (or risk aversion). It may be hard to justify much attainment performance until the upside risk is shown as a *CR* member:

$T^3 = <$
\# *milestones*
$<$
$<$-1 dollar, 0.1%-chance-at-$1000, expected-99.5-cent-win> @tomorrow,
$<$-1 dollar, 0.1%-chance-at-$1000, expected-99.5-cent-win> @next-week,
$<$-1 dollar, 0.1%-chance-at-$1000, expected-99.5-cent-win> @next-week-and-a-day
$>$,
\# *no momenta* $< >$,
\# *prospects* $< <1, 1, 1>, <1, 1, 1>, <.999, 0, 0> >$,
\# *investments* { buy-ticket@tomorrow },
\# *events-responses* { $<$win-lottery, spend-$600-after-tax-windfall-irresponsibly-during-three-year-payout$>$ }
$>$

Now an argument about mitigation acceptability, where the mitigation is an upside hazard in exchange for a mediocre attainment, might be produced. Presumably, the alternative is an envisionment of the status quo, though even that envisionment gives a place for the time value of money at 5%, and the opportunity cost of $1 cash on hand:

$T^4 = <$
milestones
$<$
<0 dollars difference$>$@tomorrow,
<0.001 dollars difference$>$@next-week-and-a-day
$>$,
momenta $< <5.2\%$ per year interest$>$, $<5.2\%$ per year interest$> >$,
prospects
$< <1>, <.95> >$,
investments $\{$ leave-dollar-in-bank@tomorrow $\}$,
events-responses $\{ <$rare-ten-for-one-dollar-menu-sale, spend-1> \}$
$>$

8.2 A Different Look at the Risk Analysis Example

The classic risk analysis process does not look so different in this framework. Here is the first move in the decision to build the Clinton, IL nuclear power plant for $4B with some operating costs, potential cost overrun, and CO2 impact represented:

$T^1 = <$
milestones
$<$
$<$$4 billion dollar or less cost, 9250 Gwh/yr, 0 ton or more CO2 reduction$>$@1985,
$<$$54 billion dollar or less cost, 9250 Gwh/yr, 4G ton or more CO2 reduction$>$@2035
$>$,
momenta
$<$
$<$$1B/yr, 0, 80M ton CO2 reduction/yr$>$,
$<$$1B/yr, 0, 80M ton CO2 reduction/yr$>$
$>$,
prospects
$<$
$<.75, .95, 1>$,
$<.85, .9, .9>$
$>$,
investments
$\{$ build-GE-BWR-at-Clinton-IL@1975 $\}$,
no events-responses $\{ \}$
$>$

At least three events are immediately contemplated: *100%-or-more-cost-overrun*, *wind-energy-cost-competitive*, and *nuclear-meltdown*. The contingent commitments are: *raise-state-taxes-1%*, *close-plant-early*, and *evacuate-Springfield-Urbana-Normal-and-Peoria*.

The adequacy of each response must take into account the relative likelihoods as well as the severity-post-mitigation. The tax-increase scenario might meet a *business-as-usual-for-IL* standard, but might actually be probable enough to raise an expected-cost dimension, which could be added to *A-space* in addition to the cost bound. The early plant closing, while improbable, might drive down the prospects of reaching the fifty-year operating life; the prospect of 9.25 Gwh production at 2035 could be counterargued, at least forcing reference to more specific classes, *e.g.*, power technologies unchallenged for 50 years, yielding a weaker probability argument. The evacuation plan might be unacceptable, and force the proposal of a different mitigation plan, such as a different location or a different reactor design.

9 Additional Discussion of Elements III: Criticism

One obvious criticism is that milestones may not be achieved, even when their prospects are quite good. This corresponds to the "good decision, poor outcome" response in decision theory. Sometimes deviation from milestone values is a mitigable event for which there is a committed response. In other cases, failure to meet the milestones may simply be an opportunity to re-plan.

Another obvious criticism is that the requirement of iterative improvement may be too strong, leading to local maxima, with sensitivity to starting proposals of trajectory. There is certainly a legitimacy to consideration of multiple trajectories, which are non-comparable with respect to the many performance measures in play. One way of using multiple maxima would be to insist that the chosen trajectory be one of the maxima, that is, a non-dominated solution. Then iterative improvement could be replaced with iterative construction of non-dominated alternative trajectories. They could also be generated with Monte Carlo starts and monotonic ascent. The problem with a reluctance to build linear orders is that a plethora of maxima can appear. The serendipity of a specific starting point solves that problem, but does so in an unsatisfying way.

It seems fair to assume, however, that the choice problem begins with aspiration levels (minimum attainments), standards of justification and standards of mitigation, and permissible expenditures (maximum costs). If we seek to put a man on the moon within ten years with high probability, typical risk for experimental pilots, and a 4% of federal budget spending limit, the counter proposal of putting twice as many men on the moon, within eight years, with added risk, and 6% of budget is perhaps not very interesting. Finding alternative maxima does not actually produce indecision if there is a trajectory that meets specification.

The search is to produce a more detailed refinement of the trajectory into specific milestones (thus improving prospects and refining investment commitments), a higher standard of risk mitigation, a longer trajectory perhaps, and alternative commitments that might achieve other desired performance. Increasing levels of attainment is possibly the last place to search for alternatives, if it concomitantly adds cost, impoverishes prospect, and lowers standards.

It is true that failure to find an acceptable trajectory, at willing cost and standard, will cause reductions in aspiration, and at this point multiple maxima can appear. Then it may be better to consult the dynamics of which aspirations are most adjustable rather than to generate mathematically unorderable alternatives willy-nilly. Failing to find an acceptable program development plan for a 1969 moon landing, the first question should be "which aspiration, or standard, should be first to yield?" The alternative question, "Do you prefer 11 years at 3% federal budget with 10% fewer mitigated hazards, or 9 years at 5% federal budget with 20% more hazards, most of which are mitigated?" seems to be an unhelpful confusion of detail rather than a commanding of goals.

Another criticism might be that probabilities and expectations are being calculated, but they are being hidden in the description of attainments and the evaluation of justifying arguments. Meanwhile, estimating the prospects of hitting milestones is just as burdensome an estimation of probability.

First, there is a lot of estimation of probabilities in the examples, and the argumentative basis for these estimates is not being given in detail. Partly this is because the examples are trying to illustrate sample argument moves. If there is insufficient description of data, there cannot be much nuanced probability argument. On the other hand, if we started with a richly described data set, such as a pharmaceutical or hospital outcome, or a baseball team's stats late in the year, then even hazards such as *sepsis*, *allergic-reaction*, *hit-by-pitch* and *pitcher-injures-arm* could be calculated without conjecture.

Furthermore, the examples here give numerical point-valued prospects to build a bridge to decision theory. In real examples, interval-valued and discrete-valued probabilities are more likely to appear. Argument permits continuing to work with such inexact values. While there may be gaps in rules for constructing arguments, there may be many hypothetical cases added, on demand, to the case base. Judgment of novel and hard cases is provided by the analyst if necessary, and these serve as bounding constraints on future cases.

Finally, with a standard in mind, such as a specific regulation, all of the numerical estimates of prospect could be replaced with the simple and easy *meets-required-standard*.

Perhaps the most important criticism to consider is whether expenditures can be hidden in the investments and responses. One would expect quantified *environmental-impact* to appear as an attribute rather than a source of investment: <*running-out-of-gas, frack-more*> is not the preferred way to justify hitting milestones. One can still hide aspects of action and consequence from the performance appraisal, but the excuse for doing so would not be "because it is hard to quantify, difficult to convert to utility measure, or impossible to ascertain at the given horizon." Meta-argument over model formulation is also possible.

Aspiration levels are defined so that vague, hard to measure attributes can be included. There could be an attribute for *effort, inconvenience, attention, diligence,* or *vigilance*. It seems that enlargement of *A-space* might be part of the dynamics of modeling not considered here. As events are considered during counterargument, some responses raise the prospect of accounting for expenditures as new dimensions in milestones.

Perhaps there are monetary investments envisioned, but there is no corresponding milestone attribute accounting for cash outlays. This could be a modeling error or weakness. But even if these aspects are not reflected in *A-space*, they still appear in the prospects and $I^\dagger$ performance measures. Philosophically, one could want the personal sacrifices to be investments rather than attainments, if only because they enable path connectedness.

If an investment requires skill, invention, magic, or miracle, situational entailment is subject to frequency counterargument as well as causal counterargument.

If a mitigation response requires redundant infrastructure, one would usually want it to be reflected as a cost attribute. But sometimes not. What if a laptop really is an extra resource on hand, that can be allocated to the project with no opportunity cost? What if an army has pre-allocated man-power, or a farm has excess, unmarketable, crop yield not written under contract? What if a gardener is willing to dig a better hole to improve the prospects of survival or a parent is willing to read another book to a child to better ensure early literacy? Perhaps a strong national effort makes a nation stronger and a strong personal effort makes a person better. During a coronary artery bypass graft operation, backup power is routinely at ready against a shared risk. Should all these things be measurable as incurred costs? Perhaps some costs are better treated like standards of proof and standards of mitigation; they are standards of feasibility, or "the cost of doing business," and they function as constraints, not as first-class objectives themselves.

A final criticism to be considered here is that the argument forms given in the examples can be quite complex and may be difficult to represent formally. After all, the advance in argument theory that makes this framework possible is a formal accounting of argument forms. This paper acknowledges that there may be a gap between argument formalism and arguments hypothesized here. There might be a small gap, or a large gap, and further investigation will reveal what work needs to be done in argument and practical reasoning to bridge the gap.

9.1 Invasion Example

$T^0 = \;<$
  # milestones
  $<$
  $<$Saddam-in-power-9, air-defenses-intact, 0%-land-control, 9-resistance, 0-losses$>$@March19,
  $<$Saddam-in-power-6, air-defenses-disrupted, 0%-land-control, 9-resistance, 0-losses$>$
                                                                                                    @March20,
  $<$Saddam-in-power-5, air-defenses-decimated, 5%-land-control, 9-resistance, 0-losses$>$
                                                                                                                @March22,
  $<$Saddam-in-power-4, air-defenses-decimated, 15%-land-control, 7-resistance, 0-losses$>$
                                                                                                                @March24,
  $<$Saddam-in-power-0, air-defenses-decimated, 95%-land-control, 1-resistance, 2-losses$>$
                                                                                                                   @April15
  $>$,
  # momenta
  $<$
  $<$0, 0, 0, 0, 0$>$,
  $<$-1, -1, 0, 0, 0$>$,
  $<$-1, 0, +10%, -2, .1$>$,
  $<$-1, 0, +10%, -2, .1$>$,
  $>$,
  # prospects
  $<$
  $<$1, 1, 1, 1, 1$>$@March19,
  $<$.8, .9, 1, 1, .9$>$@March20,
  $<$.8, .8, .9, .9, .6$>$@March22,
  $<$.7, .9, .9, .9, .6$>$@March24,
  $<$.6, .95, .5, .5, .5$>$@April15
  $>$,
  # investments
  {
  send-20-SL-cruise-missiles@March20, special-ops@March20,
                                     airstrikes@March22, Airborne101-enter-Basra@March22
  },
  # events-responses
  {
  $<$SL-missiles-miss, send-20-AL-cruise-missiles$>$,
  $<$US-aircraft-shot-down, ?$>$
  }
$>$

This first trajectory has some problems:

Entering only through Basra, the rate of reduction of Saddam in power (possibly this is even a probability, so the prospect is a meta-probability) may be optimistic, as reflected in the .8, .7, and .6 probabilities.

Entering Basra on the same day as airstrikes may put the probability of *0-losses* as low as .6. With no other investments, the milestone on *April15* is not impressively probable.

And there is no mitigation policy for the event of a lost attacking airplane. There is only a low standard to which this trajectory can be justified.

One alternative is to lower the aspiration of *0-losses* on *March24*, begin the airstrikes earlier on *March22* (increasing the probability of a US aircraft being shot down), push back the milestones and the action to *enter-Basra* on *March22*, or provide additional investments, such as a larger cruise missile strike.

There should also be a response policy for the event of a lost aircraft, or an argument (from precedent?) that such a hazard is acceptable for this plan. The obvious trajectory extension, a glaring omission of the current plan, and a desirable result of improved analysis, is a milestone that describes the situation after the removal of Saddam.

10 Conclusion

This author can actually remember when the decision models of Howard Raiffa were first gaining popularity in his management school [59], at the same time that Richard Wilson was first talking about risk analysis applied to the adoption of new technologies by the nation [67]. The two views were emerging on the same campus, both taught in the same building, Pierce Hall, and they were not yet canon in their respective choirs. Raiffa's early coauthor, John W. Pratt [57] surprised us when he said that the decision trees were not supposed to be calculating devices that produced surprising answers. They were not oracles, or crystal balls, from which the hidden solution would tumble. They were supposed to raise "What if?" questions and help structure those questions. The key word in "decision analysis" was the "analysis." This turns out to be a better description of current risk management processes than current expected utility analysis.

Decision analysis starts with a set of choices and tries to maximize expected value as a proxy for direct preference. Sometimes it is easier to start with preference, or aspirations, and determine choices. Starting with broad and sometimes vague specifications, one can search to find choices that satisfy those specifications. Risk analysis and AI planning start here. There is a search for a plan and a search for a justification of the plan. With defeasible reasoning, critical appraisal might instead be focused on the processes of search and justification. This paper relies on formal models of argumentation for that justification, while search is left as an unconstrained process (perhaps search is optimal in some way; perhaps it is serendipitous; perhaps it reflects a person's willingness to make compromises).

Argumentation is also used to manage an irreducible, partially specified, variable multi-attribute decision model with longer trajectories that might have momentum (default change of state). Argumentation also carries the weight of hard-to-quantify and hard-to-know discrete and binary attributes, and the possibility of comparing across multiple scales of attainments. Hazards and mitigations are considered iteratively and impact the justification.

The principal contribution is the mathematical framework that provides a place for formal argument in the evaluation of paths and entailments; probability is used as a standard for path coherence (not as a discount for uncertain outcome values); actions and mitigations are used to impact probability arguments, and they are used to determine standards of justifiability. Accounting is not based on expectation, but on meeting attainment levels, subject to commitment. Choice is not forced by optimal achievement, but is justified by envisionment, which includes meeting and raising standards, lengthening paths, and providing more specificity about milestones. The framework proposes navigating an increasingly specified and plausible, defensible path into the future, possibly deflected, but set aright.

Instead of maximizing lottery-based, expected-utility bundles for selecting optimal choice under precisely measurable outcomes, consider arguing milestone-based, standard-attaining trajectories for refining commitments when faced with poorly predictable hazards.

References

[1] Caroline Devred, Leila Amgoud, and Marie-Christine Lagasquie-Schiex. A constrained argumentation system for practical reasoning. *Argumentation in Multi-Agent Systems*, pages 37–56, 2009.

[2] Jean-Francois Bonnefon, Leila Amgoud, and Henri Prade. An argumentation-based approach to multiple criteria decision. *Symbolic and Quantitative Approaches to Reasoning with Uncertainty*, pages 269–280, 2005.

[3] Leila Amgoud and Henri Prade. Using arguments for making and explaining decisions. *Artificial Intelligence*, 173(3):413–436, 2009.

[4] Glasspool Anthony Cox. What's wrong with risk matrices. *Risk Analysis*, 28(2):497–512, 2008.

[5] Kevin D. Ashley. *Modeling Legal Arguments: Reasoning with Cases and Hypotheticals*, 1991.

[6] Katie Atkinson. *What Should We Do?: Computational Representation of Persuasive Argument in Practical Reasoning*, 2005.

[7] Katie Atkinson and Trevor Bench-Capon. Practical reasoning as presumptive argumentation using action based alternating transition systems. *Artificial Intelligence*, 171(10):855–874, 2007.

[8] Trevor Bench-Capon, Katie Atkinson, and Peter McBurney. Computational representation of practical argument. *Synthese*, 152(2):157–206, 2006.

[9] Trevor Bench-Capon, Katie Atkinson, and Sanjay Modgil. Argumentation for decision support. 2006.

[10] Trevor Bench-Capon and Paul E. Dunne. Argumentation in artificial intelligence. *Artificial Intelligence*, 171(10-15):619–641, 2007.

[11] Elizabeth Black and Katie Atkinson. Choosing persuasive arguments for action. *The 10th International Conference on Autonomous Agents and Multiagent Systems-Volume 3*, 2011.

[12] Barry Boehm and Wilfred J. Hansen. Spiral development: Experience, principles, and refinements, CMU/SEI-2000-SR-008. 2000.

[13] Blai Bonet and Hector Geffner. Planning as heuristic search. *Artificial Intelligence*, 129(1):5–33, 2001.

[14] Ana Gabriela Maguitman, Carlos Iván Chesñevar, and Ronald Prescott Loui. Logical models of argument. *ACM Computing Surveys*, 32:337–383, 2000.

[15] Jon Doyle. A truth maintenance system. *Artificial Intelligence*, 12(3):231–272, 1979.

[16] Jon Doyle and Richmond H. Thomason. Background to qualitative decision theory. *AI Magazine*, 20(2):55, 1999.

[17] Phan Minh Dung. On the acceptability of arguments and its fundamental role in nonmonotonic reasoning, logic programming and n-person games. *Artificial Intelligence*, 77(2):321–357, 1995.

[18] Phan Minh Dung and Tran Cao Son. An argumentation-theoretic approach to reasoning with specificity. *Proceedings of the KR Conference*, 1996.

[19] Phan Minh Dung and Tran Cao Son. An argument-based approach to reasoning with specificity. *Artificial Intelligence*, 133(1):35–85, 2001.

[20] George Ferguson and James F. Allen. Arguing about plans: plan representation and reasoning for mixed-initiative planning. *Proceedings of the AIPS Conference*, 1994.

[21] Peter C. Fishburn. Mean-risk analysis with risk associated with below-target returns. *The American Economic Review*, pages 116–126, 1977.

[22] Peter C. Fishburn. *A Survey of Multiattribute/Multicriterion Evaluation Theories*, 1978.

[23] David Glasspool, Dan Grecu, Sanjay Modgil, Matthew South, and Vivek Patkar. Argumentation-based inference and decision making - a medical perspective. *IEEE Intelligent Systems*, 6:34–41, 2007.

[24] John Fox and Sanjay Modgil. From arguments to decisions: extending the toulmin view. *Arguing on the Toulmin Model*, pages 273–287, 2006.

[25] David P. Gauthier. *Practical Reasoning*, 1963.

[26] John Fox, Ayelet Oettinger, David Glasspool, and James Smith-Spark. Argumentation in decision support for medical care planning for patients and clinicians. *AAAI Spring Symposium: Argumentation for Consumers of Healthcare*, pages 58–63, 2006.

[27] Thomas F. Gordon and Douglas Walton. Proof burdens and standards. pages 239–258, 2009.

[28] Jaap Hage and Bart Verheij. Reason-based logic: A logic for reasoning with rules and reasons. *Information and Communications Technology Law*, 3(2-3):171–209, 1994.

[29] Jean-Claude Latombe, David Hsu, and Rajeev Motwani. Path planning in expansive configuration spaces. *IEEE International Conference on Robotics and Automation*, 1997.

[30] Richard C. Jeffrey. *The Logic of Decision*, 1965.

[31] N. I. Karacapilidis and Thomas Gordon. Dialectical planning. *Proceedings of the 14th IJCAI Workshop on Intelligent Manufacturing Systems, Montreal*, 1995.

[32] Petr Svestka, Jean-Claude Latombe, Lydia E. Kavraki, and Mark H. Overmars. Probabilistic roadmaps for path planning in high-dimensional configuration spaces. *IEEE Transactions on Robotics and Automation*, 12(4):566–580, 1996.

[33] Ralph L. Keeney and Raiffa Howard. *Decisions with Multiple Objectives: Preferences and Value Tradeoffs*, 1976.

[34] Tim Kelly and John McDermid. Safety case patterns-reusing successful arguments. *IEEE Colloquium on Understanding Patterns and their Application to Systems Engineering*, pages 3–3, 1998.

[35] John Fox, Paul Krause, and Philip Judson. An argumentation-based approach to risk assesment. *IMA Journal of Management Mathematics*, 5(1):249–263, 1993.

[36] John Fox, Paul Krause, and Philip Judson. Is there a role for qualitative risk assessment. *Proceedings of the Eleventh Conference on Uncertainty in Artificial Intelligence*, 1995.

[37] John Fox, Philip Judson, Paul Krause, and Mukesh Patel. Qualitative risk assessment fulfils a need. *Applications of Uncertainty Formalisms*, pages 138–156, 1998.

[38] Henry E. Kyburg Jr. The reference class. *Philosophy of Science*, pages 374–397, 1983.

[39] John Lintner. The valuation of risk assets and the selection of risky investments in stock portfolios and capital budgets. *The Review of Economics and Statistics*, pages 13–37, 1965.

[40] Ronald P. Loui. Defeasible decisions. *Proceedings of the Fifth Uncertainty in AI Conference*, 1989.

[41] Ronald P. Loui. Two heuristic functions for decision. *Proceedings of the AAAI Spring Symposium*, 1989.

[42] Ronald P. Loui. Defeasible specification of utilities. *Knowledge Representation and Defeasible Reasoning*, 1990.

[43] Ronald P. Loui. Scientific and legal theory formation in an era of machine learning: remembering background rules, coherence, and cogency in induction. *APA Computers and Philosophy Newsletter*, 2014.

[44] Ronald P. Loui and Jeff Norman. Rationales and argument moves. *Artificial Intelligence and Law*, 3(3):159–189, 1995.

[45] Ronald P. Loui. Process and policy: resource-bounded non-demonstrative reasoning. *Computational Intelligence*, 14(1):1–38, 1998.

[46] Peter McBurney and Simon Parsons. Truth or consequences: using argumentation to reason about risk. *BPS Symposium on Practical Reasoning*, 21, 1999.

[47] Peter McBurney and Simon Parsons. Risk agoras: Dialectical argumentation for scientific reasoning. *Proceedings of the Sixteenth Conference on Uncertainty in Artificial Intelligence*, 2000.

[48] Peter McBurney and Simon Parsons. Determining error bounds for hypothesis tests in risk assessment: a research agenda. *Law, Probability and Risk*, 1(1):17–36, 2002.

[49] Juan Carlos Nieves and Roberto Confalonieri. A possibilistic argumentation decision making framework with default reasoning. *Fundamenta Informaticae*, 113(1):41–61, 2011.

[50] John L. Pollock. Defeasible reasoning. *Cognitive Science*, 11(4):481–518, 1987.

[51] John L. Pollock. *Nomic Probability and the Foundations of Induction*, 1990.

[52] Dan Ionita, Henry Prakken, and Roel Wieringa. Risk assessment as an argumentation game. *Computational Logic in Multi-agent Systems*, pages 357–373, 2013.

[53] Henry Prakken and Giovanni Sartor. *Logical Models of Legal Argumentation*, 1997.

[54] Henry Prakken and Giovanni Sartor. Modelling reasoning with precedents in a formal dialogue game. *Judicial Applications of Artificial Intelligence*, pages 127–183, 1998.

[55] Henry Prakken and Giovanni Sartor. On modelling burdens and standards of proof in structured argumentation. *JURIX*, 2011.

[56] Henry Prakken and Gerard Vreeswijk. Logics for defeasible argumentation. *Handbook of Philosophical Logic*, pages 219–318, 2002.

[57] Howard Raiffa, John W. Pratt, and Robert Schlaifer. The foundations of decision under uncertainty: An elementary exposition. *Journal of the American Statistical Association*, 59(306):353–375, 1964.

[58] Iyad Rahwan and Leila Amgoud. An argumentation based approach for practical reasoning. *Proceedings of the Fifth International Joint Conference on Autonomous Agents and Multiagent Systems*, 2006.

[59] Howard Raiffa. *Applied Statistical Decision Theory*, 1974.

[60] Nicholas Rescher. *Dialectics: A Controversy-Oriented Approach to The Theory of Knowledge*, 1977.

[61] Leonard J. Savage. *The Foundations of Statistics*, 1954.

[62] Glenn Shafer. Savage revisited. *Statistical Science*, pages 463–485, 1986.

[63] Blake Shepard, Cynthia Matuszek, C. Bruce Fraser, Wiliam Wechtenhiser, David Crabbe, Zelal Güngördü, John Jantos, Todd Hughes, Larry Lefkowitz, Michael Witbrock, Doug Lenat, and Erik Larson. A knowledge-based approach to network security: applying Cyc in the domain of network risk assessment. *Proceedings of the National Conference on Artificial Intelligence*, 2005.

[64] Guillermo R. Simari and Ronald P. Loui. A mathematical treatment of defeasible reasoning and its implementation. *Artificial Intelligence*, 53(2):125–157, 1992.

[65] Katie Atkinson, Peter McBurney, Sanjay Modgil, Pancho Tolchinsky, and Ulises Cortés. Agents deliberating over action proposals using the proclaim model. *International Central and Eastern European Conference on Multi-Agent Systems*, 2007.

[66] Frank Dignum, J-J. Ch Meyer, Henry Prakken, Thomas L. van der Weide, and G. A. W. Vreeswijk. Practical reasoning using values. *Argumentation in Multi-Agent Systems*, pages 79–93, 2009.

[67] Andrew J. Van Horn and Richard Wilson. *Status of Risk-Benefit Analysis. No. BNL-22282*, 1976.

[68] Koen V. Hindriks, Wietske Visser, and Catholijn M. Jonker. An argumentation framework for deriving qualitative risk sensitive preferences. *Modern Approaches in Applied Intelligence*, 2011.

[69] Gerard A. W. Vreeswijk. Abstract argumentation systems. *Artificial Intelligence*, 90(1):225–279, 1997.

[70] Michael P. Wellman and Mark Derthick. *Formulation of Tradeoffs in Planning under Uncertainty*, 1990.

Introducing Bayesian Argumentation Networks

D. M. Gabbay
Department of Informatics, King's College London,
Ashkelon Academic College, Israel,
Bar Ilan University, Ramat Gan, Israel
University of Luxembourg, Luxembourg.
http://www.inf.kcl.ac.uk/staff/dg
dov.gabbay@kcl.ac.uk

O. Rodrigues
Department of Informatics, King's College London, The Strand, London, WC2R 2LS, UK,
http://www.inf.kcl.ac.uk/staff/odinaldo
odinaldo.rodrigues@kcl.ac.uk

Abstract

We give a faithful interpretation of Bayesian networks into a version of numerical argumentation networks based on Łukasiewicz infinite-valued logic with product conjunction. The advantages of such a translation, beyond the theoretical aspects of it, are hopefully threefold: 1) importing updating algorithms into argumentation networks; 2) importing the handling of loops into cyclic Bayesian networks; and 3) importing logical proof theory into Bayesian networks.

Keywords: Bayesian Networks, Argumentation Theory, Numerical Networks

1 Introduction

In this paper, we compare probabilistic argumentation with Bayesian networks and motivate the new definition of Bayesian Argumentation Networks. We examine what extra features are needed to extend traditional abstract argumentation frameworks to enable the extended frameworks to simulate Bayesian networks. Once we identify such features, then we can call the extended argumentation frameworks by the name

Bayesian Argumentation Networks. We shall see later that the extra features are all well-known features existing in the literature in various contexts.

In order to illustrate these ideas, consider the network $\langle S, R \rangle$ of Fig. 1. In this figure, the arrows just indicate parenthood. If the arrows are considered as attacks, then $\langle S, R \rangle$ is a traditional abstract argumentation framework (henceforth a "Dung network"), and there is only one complete extension $E = \{X, Y, U, W\}$ (with $A =$ "out").

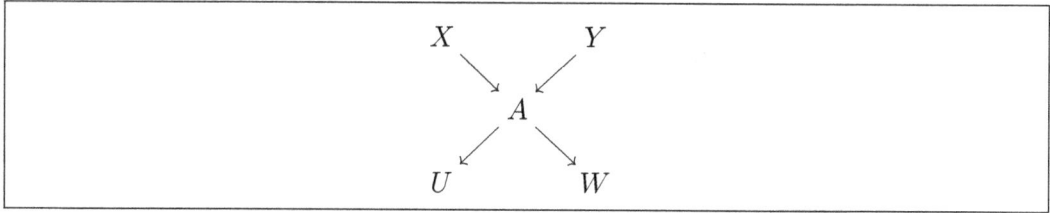

Figure 1: A sample argumentation network.

The operating assumptions (which are violated in Bayesian networks) are:

1. Since there is no connection (i.e., attack) going into X and into Y, then X and Y are "in" (intuitively meaning $X = Y = \top$).

2. Since U and V have the same parent A, we treat U and V in the same way.

3. We do not mind having cycles, i.e., R need not be acyclic.

There are other assumptions in the case of argumentation networks, but let us concentrate only on the ones above.

Bayesian networks do not allow for cycles (R must be acyclic) and they do not determine the values of nodes without parents, such as X and Y. Moreover, they are not committed to treating nodes with the same parents (such as U and W) the same way. Such a view is not new to argumentation. In fact such a view is shared by Abstract Dialectical Frameworks (ADFs) [6]. In an ADF, each node α depends on its parents, say $\{\beta_1, \ldots, \beta_n\}$ via a Boolean formula Ψ_α specific to α. Thus, we have that α depends on $\Psi_\alpha(\beta_1, \ldots, \beta_n)$ in the ADF case and we want that

$$\alpha \leftrightarrow \Psi_\alpha(\beta_1, \ldots, \beta_n).$$

In Dung's networks, the same constraint Ψ_α is imposed on all nodes α, namely

$$\alpha \leftrightarrow \bigwedge_{i=1}^{n} \neg \beta_i$$

where "∧¬" is the Peirce-Quine connective

$$\downarrow \{E_i\} = \wedge_{i=1}^{n} \neg E_i^{\,1}$$

Still, we do not have many options when we treat the network of Fig. 1 as an ADF. Since X and Y depend on the empty set, then each can be either $\top$ or $\bot$. A depends on X and Y, and these being $\top$ or $\bot$ allow for A to be either $\top$ or $\bot$ and similarly for U and W. So depending on the Boolean functions Ψ_α employed, we can get all possible distributions of $\{\bot, \top\}$ among the nodes in Fig. 1.[2]

If we allow source nodes such as X and Y to have arbitrary given values in $[0, 1]$ and are able to describe the desired dependencies between node values in an argumentation context, then we can bridge the gap between the Bayesian and argumentation representations and hence analyse the properties of the former under the perspective of the latter. This can bring benefits to both areas which we will discuss later.

We find that the probabilistic approach to argumentation is the nearest we can get to Bayesian networks. We identify that what is missing in the probabilistic approach is a representation of conditional probabilities, a feature which is central in Bayesian networks. We further realise that if we define new argumentation networks based on joint attacks defined numerically using Łukasiewicz infinite-valued logics, we will have what we need. The integration of conditional probabilities and joint attacks is one of the objectives of this paper.

The rest of the paper is structured as follows. We start with a description of the probabilistic approach to argumentation in Section 2. The section is written in a Socratic manner, leading the reader to our conclusions using examples and semi-formal definitions. Section 3 gives the formal definitions in a systematic manner. Section 4 contains a comprehensive example illustrating what we have done. Section 5 deals with complexity issues. Section 6 discusses related literature [3, 4, 9, 16, 5, 19, 20, 8, 21, 22, 24, 25, 27]. In Section 7, we conclude with a discussion and directions for future research.

---

[1] Notice that the other boolean connectives can be defined in terms of $\downarrow$, for instance, $\neg P \equiv P \downarrow P$.

[2] As pointed out by one of the referees, the reader might think that this is a shortcoming of ADFs in the sense that initial arguments, being dependent on the empty set, can only have a fixed value. However, there is also the possibility to consider all initial arguments A as self-looping with acceptance condition A. In this case most semantics then yield a "guessing" value for A. We should however be cautious in not allowing too many modifications. It is known that enough modifications can reduce ADFs to traditional argumentation systems (see [14]).

2 Background Discussion

This section develops in a Socratic manner the features we need to reach a proper representation of a Bayesian network as an extension of an argumentation network. We therefore turn to the probabilistic approach to argumentation, it being the nearest to Bayesian networks. We need some notation before we describe it. As a starting point, we consider the elements of $S = \{X, Y, A, U, V\}$ as classical propositional atoms capable of getting the values $\alpha = \top$ (corresponding to α is "in") and $\alpha = \bot$ (corresponding to α is "out"). Let us understand the term *full conjunction of literals* to mean a conjunction containing for each atom α of the language (which is assumed to be finite) either α or $\neg \alpha$. Any full conjunction of literals of the form

$$e = \wedge_i \alpha_i^{\pm}$$

can be considered a classical model $m(e)$. We have

$$m(e) \models \alpha \text{ iff } e \vdash \alpha$$

We can also associate with e a subset S_e of S, $S_e = \{\alpha \in S \mid e \vdash \alpha\}$ (remember that the elements of S are atoms without negation). So if we assign probability distributions π on the models $m(e)$ or on the set of full conjunctions of literals $\{e\}$, we get a traditional probability function π on the space $\Omega = 2^{2^S}$ = families of models = $2^{\{e\}}$ = the set of all propositional well-formed formulae (wffs) built-up from the atoms of S.

In our example, $S = \{X, Y, A, U, V\}$. 2^S = all subsets of S = all models of the language S. $\Omega = 2^{2^S}$ = all possible sets of models. So π gives a value $0 \leq \pi(m) \leq 1$, for each model m of S. We have that $\sum_m \pi(m) = 1$.

According to [13], the probability $\pi(m)$ for a typical conjunctive model $m(e)$ where $e = \wedge_{\alpha \in S} \alpha^{\pm}$ can be given in two main ways.

i) The semantic way, which gives values $\pi(m(e))$ directly for each e.

ii) The syntactic way, which gives values $\pi(\alpha)$, for each $\alpha \in S$, and then $\pi(m(e))$ is defined as the product

$$\prod_\alpha \pi^{\pm}(\alpha)$$

where $\pi^+(\alpha) = \pi(\alpha)$ and $\pi^-(\alpha) = 1 - \pi(\alpha)$.

So for example in Fig. 1, we either give probability directly to each model, e.g., to $e = X \wedge Y \wedge \neg A \wedge U \wedge \neg W$ or give probabilities to each of X, Y, A, U and W, and then the probability of **e** can be calculated as

$$\pi(X) \cdot \pi(Y) \cdot (1 - \pi(A)) \cdot \pi(U) \cdot (1 - \pi(W))^3$$

The version of the probabilistic approach to both Dung or ADF which can be compared with the Bayesian approach is the syntactical one, *ii)* above, the one which assigns probabilities to nodes (not the one that assigns probabilities to the subnetworks). Thus each of the nodes X, Y, A, U and V is assigned a probability value.

The most general case of getting such probability is to regard the arguments as atoms in a space (i.e., $S = \{X, Y, A, U, V\}$) and assign probabilities to the subsets of S. This is a traditional probability distribution. Note that the subsets $E \subseteq \Omega = 2^{2^S}$ can also be identified with sets of models of formulas built-up using atoms from S.

We now show a connection of probabilistic argumentation with Bayesian argumentation. Both Dung and ADFs would read Fig. 1 as follows. The figure suggests the probability space being the family $\Omega = 2^{2^S}$ of all subsets of $S = \{X, Y, A, U, V\}$ and the connections (arrows) in the figure suggest restriction on the probability π on Ω. We want to consider only those probabilities which satisfy for every $\alpha \in S$ with parents $\{\beta_1, \ldots, \beta_n\}$ the following:

$$\pi(\alpha) = \pi(\psi_\alpha(\beta_1, \ldots, \beta_n))$$

Remember that each wff defines a set of models in which it holds, and π is a probability on sets of models. Thus π gives a number $0 \leq \pi(m) \leq 1$ to each model m of the language of S with $\sum_m \pi(m) = 1$.

Bayesian argumentation looks at the elements of S as random variables capable of getting $\top$ or $\bot$, and regards all probability functions $P(\alpha_1, \ldots, \alpha_n)$ where $\{\alpha_i\} = S$.

In our case we have probability functions $\boldsymbol{P}(X, Y, A, U, V)$. Thus for each combination of values of $\top$ or $\bot$ to the variables in S, $\boldsymbol{P}$ will give a probability.

Such a combination can also be viewed as a model m for the language of S, and so $\boldsymbol{P}$ gives probability to models. This is the same as π, but the restrictions on $\boldsymbol{P}$ and the manipulation of $\boldsymbol{P}$ are different in this case. We have, according to the Bayesian view, that Fig. 1 gives the dependencies of the variables on each other. Let $\{\beta_1, \ldots, \beta_n\}$ be all the parents of α and $\boldsymbol{P}(\alpha|Z)$ denote the conditional probability of α given Z. We have the following equations:

$$\boldsymbol{P}(\alpha) = \boldsymbol{P}(\alpha \mid \beta_1, \ldots, \beta_n) \cdot \boldsymbol{P}(\beta_1, \ldots, \beta_n)$$

---

[3] See [13].

If α has no parents, then $\boldsymbol{P}(\alpha)$ must be given. If α does have parents $\{\beta_1,\ldots,\beta_n\}$, then $\boldsymbol{P}(\alpha \mid \beta_1,\ldots,\beta_n)$ must be given. $\boldsymbol{P}(\alpha \mid \beta_1,\ldots,\beta_n)$ can be given as a function giving a value in $[0,1]$ for every choice of $\top, \bot$ to each β_i.

Thus the Bayesian approach specifies a syntactical type probability on the atoms $\alpha \in S$, by using the graph of the network and giving conditional probabilities for the dependencies of the graph.

So for the network of Fig. 1 we need the following values to specify a specific Bayesian distribution $\boldsymbol{P}$:

- Values of the probabilities of the source nodes X and Y, i.e., $\boldsymbol{P}(X)$ and $\boldsymbol{P}(Y)$.

- A table of values v, describing the coefficients of the function $\boldsymbol{P}(A|X,Y)$. We use the notation $F_{A|11}$ for the case $A|X \wedge Y$, $F_{A|10}$ for the case $A|X \wedge \neg Y$, etc. We denote the transmission coefficient for each case as $e_{F_{A|11}}$, $e_{F_{A|10}}$, and so on:

X	Y	v
$\top$	$\top$	$e_{F_{A|11}}$
$\top$	$\bot$	$e_{F_{A|10}}$
$\bot$	$\top$	$e_{F_{A|01}}$
$\bot$	$\bot$	$e_{F_{A|00}}$

- A table for $\boldsymbol{P}(U|A)$:

A	v
$\top$	$e_{F_{U|1}}$
$\bot$	$e_{F_{U|0}}$

- A table for $\boldsymbol{P}(W|A)$:

A	v
$\top$	$e_{F_{W|1}}$
$\bot$	$e_{F_{W|0}}$

Note that the network of Fig. 1 actually depicts Pearl's famous Earthquake example [20] as described in the book "Bayesian Artificial Intelligence" by Korb and Nicholson [17]:

"You have a new burglar alarm installed. It reliably detects burglary, but also responds to minor earthquakes. Two neighbors, John and Mary, promise to call the police when they hear the alarm. John always calls when he hears the alarm, but sometimes confuses the alarm with the phone ringing and calls then also. On the other hand, Mary likes loud music and sometimes doesn't hear the alarm. Given evidence about who has and hasn't called, you'd like to estimate the probability of a burglary."

Replacing X with "Burglary", Y with "Earthquake", A with "Alarm", U with "John calls", and W with "Mary calls", gives the Bayesian network:

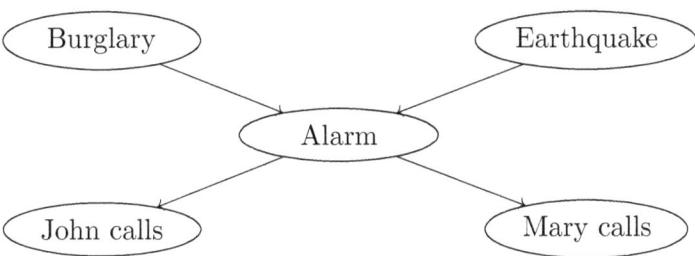

For simplicity, we will continue to use the letters X, Y, A, U and W.

Let us assume the following values. $\boldsymbol{P}(X) = 0.01$, giving $\boldsymbol{P}(X = \top) = 0.01$, and $\boldsymbol{P}(X = \bot) = 0.99$; $\boldsymbol{P}(Y = \top) = 0.02$, giving $\boldsymbol{P}(Y = \bot) = 0.98$; and the values given by the tables below:

X	Y	v
$\top$	$\top$	$e_{F_{A|11}} = 0.95$
$\top$	$\bot$	$e_{F_{A|10}} = 0.94$
$\bot$	$\top$	$e_{F_{A|01}} = 0.29$
$\bot$	$\bot$	$e_{F_{A|00}} = 0.001$

A	v
$\top$	$e_{F_{U|1}} = 0.9$
$\bot$	$e_{F_{U|0}} = 0.05$

A	v
$\top$	$e_{F_{W|1}} = 0.70$
$\bot$	$e_{F_{W|0}} = 0.01$

We can now compute the probabilities $\boldsymbol{P}(A)$, $\boldsymbol{P}(U)$ and $\boldsymbol{P}(W)$.

$$\begin{aligned}
\boldsymbol{P}(A) &= e_{F_{A|11}} \times \boldsymbol{P}(X) \times \boldsymbol{P}(Y) + \\
&\quad e_{F_{A|10}} \times \boldsymbol{P}(X) \times (1 - \boldsymbol{P}(Y)) + \\
&\quad e_{F_{A|01}} \times (1 - \boldsymbol{P}(X)) \times \boldsymbol{P}(Y) + \\
&\quad e_{F_{A|00}} \times (1 - \boldsymbol{P}(X)) \times (1 - \boldsymbol{P}(Y)). \\
\boldsymbol{P}(W) &= e_{F_{W|1}} \times \boldsymbol{P}(A) + e_{F_{W|0}} \times (1 - \boldsymbol{P}(A)). \\
\boldsymbol{P}(U) &= e_{F_{U|1}} \times \boldsymbol{P}(A) + e_{F_{U|0}} \times (1 - \boldsymbol{P}(A)).
\end{aligned}$$

So

$$\begin{aligned}
\boldsymbol{P}(A) &= 0.95 \times 0.01 \times 0.02 + \\
&\quad 0.94 \times 0.01 \times 0.98 + \\
&\quad 0.29 \times 0.99 \times 0.02 + \\
&\quad 0.001 \times 0.99 \times 0.98. \\
&= 0.00019 + 0.009212 + 0.005742 + 0.0009702 \\
&= 0.0161142 \approx 0.016
\end{aligned}$$

So $\boldsymbol{P}(\neg A) \approx 0.984$. Now for U and W, we get

$$\begin{aligned}
\boldsymbol{P}(W) &= (0.7 \times 0.016) + (0.01 \times 0.984) = 0.0112 + 0.00984 \approx 0.021. \\
\boldsymbol{P}(U) &= (0.9 \times 0.016) + (0.05 \times 0.984) = 0.014 + 0.0492 \approx 0.063.
\end{aligned}$$

Thus we can see that we have a syntactical probability distribution on S.

2.1 A Common Ground for Bayesian, Argumentation and Abstract Dialectical Frameworks

To compare Bayesian networks with say ADFs or with traditional Dung networks, we need to go to a common ground. First we note that with any formal system, whether it be a logic such as classic or intuitionistic logic, or whether it be a Bayesian, ADF or traditional argumentation network there are always two components. The first one is the intended meaning of the system. The second is the formal mathematical representation of the system and the mathematical machinery of handling it. When we compare two such systems we can compare them in regard to their formal machinery or we can compare them in their intended meaning. It may be that two systems have the same formal machineries but completely different meanings. This happens a lot in modal logics with possible world semantics. In the case of a network $\langle S, R \rangle$, the intended meaning may impose some restrictions on the graph. In an argumentation network the arrows mean attack; the variables get values "in", "out" and "undecided"; and unattacked nodes must get value "in". In Bayesian networks the arrows represent dependencies and there is the requirement of the network being acyclic. In addition, nodes with the same parents in Bayesian networks can behave differently, which is not the case in the traditional argumentation but is the case in ADFs. In ADFs the arrows represent dependencies and there is no requirement of being acyclic. Having said all that let us now compare the systems on the basis of their mathematical machinery which can be captured by the two points below.

a. Since Bayesian networks allow for points without parents to have an arbitrary probability assigned to them, Bayesian networks can agree to limit such assignment for the sake of common grounds with argumentation and assign probability 1, we can assume a similar sacrifice and the same property for ADFs. If, on the other hand, we want to leave Bayesian networks as they are (not ask them to make any limitations) but we still want to have common ground with respect to this property with ordinary Dung networks, we can modify Dung's networks, and add for each node α a new node called $\neg\alpha$, with α and $\neg\alpha$ attacking each other. This will allow any node α which was originally unattacked to get any value in the modified network, because it will be part of the cycle $\{\alpha, \neg\alpha\}$.

We can also assume, for Bayesian networks, the sacrifice limitation that nodes with the same parents behave the same (we can call these Bayesian networks *BNA nets*.[4] We can also add this requirement (that nodes with the same parents behave the same way) as an additional assumption on ADFs to make them more in line with traditional Dung networks.

b. Bayesian networks are acyclic and since ADFs and traditional Dung networks can also be acyclic, let us accept this additional restriction on them.

So we compare acyclic probabilistic ADFs with BNA nets and see what else we need to add to argumentation networks to be able to implement Bayesian networks in them.

The above discussion outlined several possibilities for finding common grounds between Bayesian networks and Dung's networks. We made what we think is the best choice/approach, which we now proceed to explain.

---

[4]The letters "NA" stand for "nice-to-argumentation".

Looking again at Fig. 1, we see that what is missing in order to do a proper comparison is the fact that Bayesian networks have the conditional probabilities. Let us look at Fig. 2, which is the top part of Fig. 1.

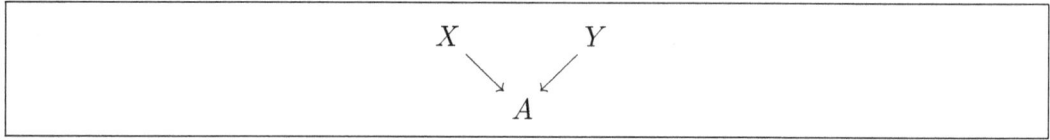

Figure 2: The top part of the network in Fig. 1.

What the Bayesian approach does is to give arbitrary values $P(X)$, $P(Y)$ (so we have syntactical probabilities for X and Y), but to get $P(A)$, it uses transmission values as given in Fig. 3. $e_{F_{A|ij}}$ are transmission coefficients in the sense of [1, 2] and F_{ij} is an attack formation in the sense of [12], to be explained below and formally defined in Section 3.

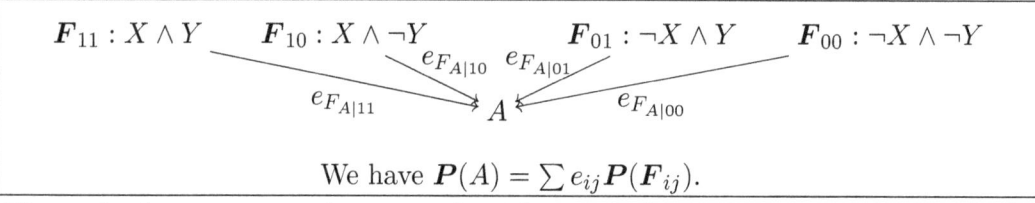

Figure 3: An argumentation network with attack formations, transmission coefficients and joint attacks.

So what we need to accommodate Bayesian networks are argumentation networks with attack formations, transmission coefficients, and joint attacks obeying the attack formula of Fig. 3. The semantics of such networks is best given using the *equational approach* [15].[5]

---

[5]There are several different interpretations of basic argumentation notions. In traditional Dung networks arcs represent attacks while in ADFs they represent dependencies It is natural then to think that ADFs are closer in meaning to Bayesian networks because certainly arcs in Bayesian networks are not attacks but dependencies. Our reader may therefore be puzzled at our translation of Bayesian networks into argumentation where arcs represent attacks. We even use joint attacks. We remind the reader that ADFs can be translated into traditional argumentation networks using joint attacks and additional nodes. The additional nodes are used to help simulate the boolean dependencies (see [14]). It may be possible to translate Bayesian networks into ADFs, but we would need additional points and some kind of fuzzy propagation. It therefore makes more sense to translate directly into traditional networks with attacks. Note also that numerical values associated with nodes can have several interpretations: 1) a fuzzy truth-value; 2) a probability value expressing uncertainty about argument acceptance; 3) a value obtained in the context of the equational

We now explain the components needed.

a. Attack formations

Consider Fig. 4 (L) and Fig. 4 (R). In Fig. 4 (L), we have that α attacks β. We have

(a) If $\alpha =$ "in", then $\beta =$ "out"

(b) If $\alpha =$ "out", then $\beta =$ "in" (unless β is attacked by something else that is not "out")

(c) If $\alpha =$ "undecided", then $\beta =$ "undecided" (unless it is attacked by something else that is "in", in which case β has to be "out")

Figure 4

In Fig. 4 (R), we have two intermediary points unique to the pair (α, β). We view this as an *attack formation*. It does the job of Fig. 4 (L). (a), (b) and (c) still hold for Fig. 4 (R). Fig. 4 (R) is a general replacement for Fig. 4 (L), used by Gabbay in [14]. It allows for the implementation of higher level attacks.[6]

---

approach as the result of some calculation. 4) a Bayesian probability value. The conditions when two of such values coincide need to be investigated. For example, we do know that for the Eq_{inv} equational approach the numerical values can be viewed as probabilistic values where the arguments are mutually independent [13].

[6]Higher level attacks are attacks on attacks. So for example, an academic professional argument β put forward in favour of promoting three members of staff to the rank of full professor by virtue of

For example the attack $\gamma \to (\alpha \to \beta)$ can be implemented as $\gamma \to Y_{\alpha,\beta}$. The two additional dummy points $X_{\alpha,\beta}$ and $Y_{\alpha,\beta}$ are just two "transmitting points", which allow the node γ to attack the transmission by attacking the point $Y_{\alpha,\beta}$.

We do not need higher level attacks in this paper but it is useful to know how useful attack formations are in translations/implementations. We use the notation $\boldsymbol{F}[\alpha, \beta]$ as in Fig. 5.

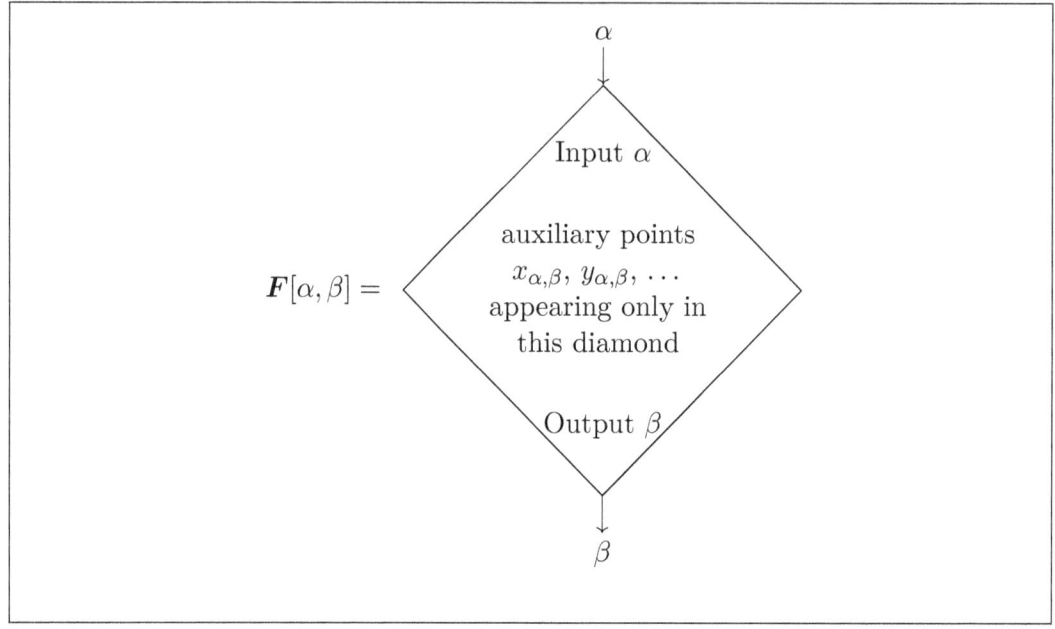

Figure 5

Attack formations can be single arguments as in Fig. 6. The argument A is both the input and the output point of the formation.

Attack formations can attack each other, as in Fig. 7. The output point of formation $\boldsymbol{F}_1$ attacks the input point of formation $\boldsymbol{F}_2$.

b. Transmission coefficients

---

their brilliant performance may be attacked by an argument α which says that there is not enough money to pay their higher salaries and benefits if indeed promoted. If the claims of both arguments are true (i.e., the individuals did perform well and indeed there is not enough money to pay for the extra expenditure caused by the promotion, there is nothing to say except to put forward an argument γ which says that budgetary considerations should not be arguments against promotion. Here γ is a higher level attack on the very attack arrow $\alpha \to \beta$. We write this as $\gamma \to (\alpha \to \beta)$.

Bayesian Argumentation Networks

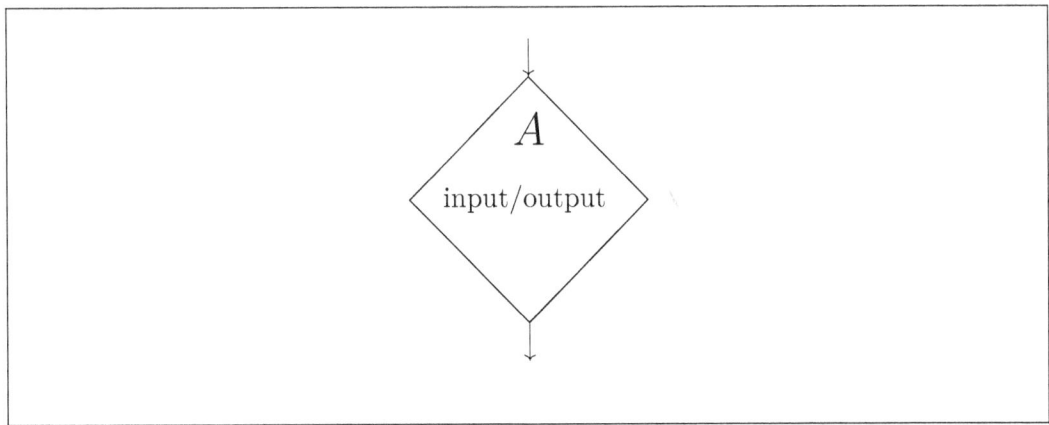

Figure 6

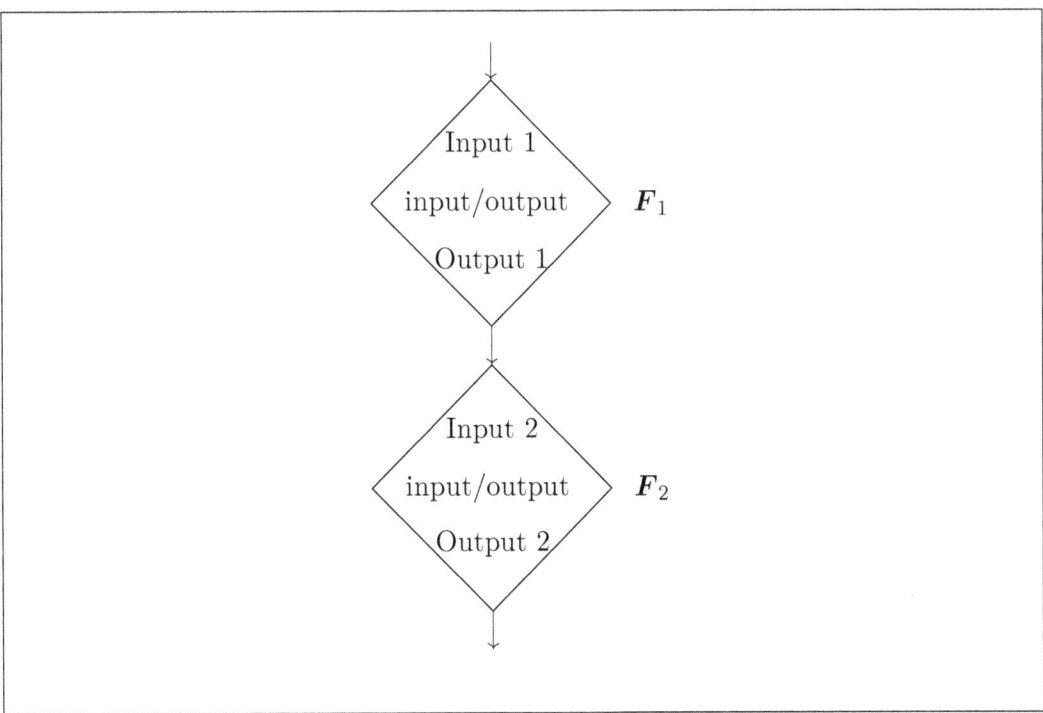

Figure 7: An attack formation attacking another attack formation.

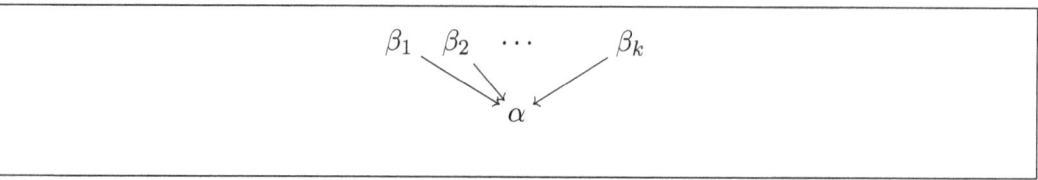

Figure 8: A typical attack configuration in an argumentation network.

Consider Fig. 8, under the equational approach of [15]. Let $\beta_1, \ldots, \beta_k$ be all of the attackers of node α. Each node has a numerical value, $f(\alpha)$, $f(\beta_1)$, ..., $f(\beta_k)$. The equational approach (under the Eq_{inv} Equational semantics) requires that

$$f(\alpha) = \prod_{i=1}^{k}(1 - f(\beta_i)) \quad (1)$$

The equational approach obtains f as a solution to the equations of type (1), for every $\alpha \in S$, and *at least* all the preferred extensions (in Dung's sense) are obtained via the correspondence:

(a) $\alpha =$ "in", if $f(\alpha) = 1$

(b) $\alpha =$ "out", if $f(\alpha) = 0$

(c) $\alpha =$ "undecided", if $0 < f(\alpha) < 1$

Such solutions also can be seen as syntactical probability distributions for the nodes as shown in [13]. So, if we implement Bayesian networks in argumentation networks with the Eq_{inv} interpretation, we hope to get the Bayesian probabilities as preferred extensions.

When we have a transmission coefficient e_i between the attacker β_i and α, the strength of the attack from β_i is adjusted by the coefficient e_i, and hence its value is only $e_i \times \beta_i$. Thus, we get for Fig. 9 of attacks with transmission coefficients that

$$f(\alpha) = \prod_{i=1}^{k}(1 - e_i \times f(\beta_i)) \quad (2)$$

It is worth noting that the transmission coefficient does not change the expressive power of Eq_{inv}, since the effects of the coefficients can be implemented through additional nodes in Eq_{inv}.

c. Joint attacks

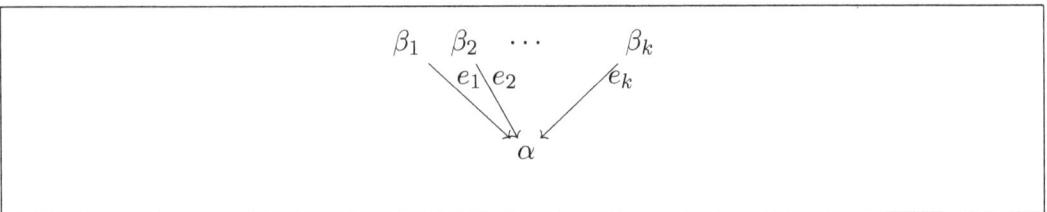

Figure 9: A typical attack configuration in an argumentation network with transmission coefficients.

In [14], Gabbay et. al. used the notation of Fig. 10 in the context of Fibring networks, joint attacks and disjunctive attacks. The idea of joint attacks on its own was earlier introduced in [18]. The intended meaning of a joint attack is that α is "out", if all β_i are "in". In a numerical context (i.e., under an equational approach), we can write the equation

$$f(\alpha) = 1 - \prod_i f(\beta_i)$$

Clearly α is "out" exactly when all of β_i are "in".

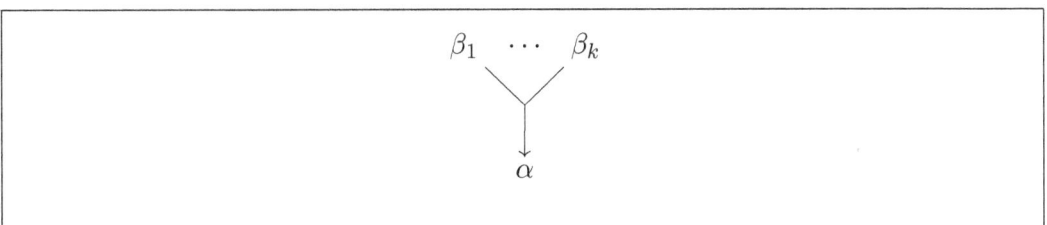

Figure 10: A joint attack from $\beta_1,\ldots,\beta_k$ to α.

Remark 1. *Note that if we have such joint attacks (as given in [18]) under the equational approach, we can implement products (π-attacks) with the help of auxiliary points. For example, the value of α as the product of β_1 and β_2 in Fig. 11 can be implemented as Fig. 12 and vice-versa. In Fig. 12 we have that $f(x) = 1 - \beta_1 \cdot \beta_2$ and $f(\alpha) = 1 - x = \beta_1 \cdot \beta_2$.*

Fig. 13 can be implemented as Fig. 14. In Fig. 13, we have that $f(\alpha) = 1 - \beta_1 \cdot \beta_2$. In Fig. 14, $f(y) = \beta_1 \cdot \beta_2$ and $f(\alpha) = 1 - y = 1 - \beta_1 \cdot \beta_2$.

For implementing Bayesian networks we need a different understanding for joint attacks, yielding a different equation. What we need is the following equation:

$$f(\alpha) = \min(1, \sum_j (1 - f(\beta_j))) \tag{3}$$

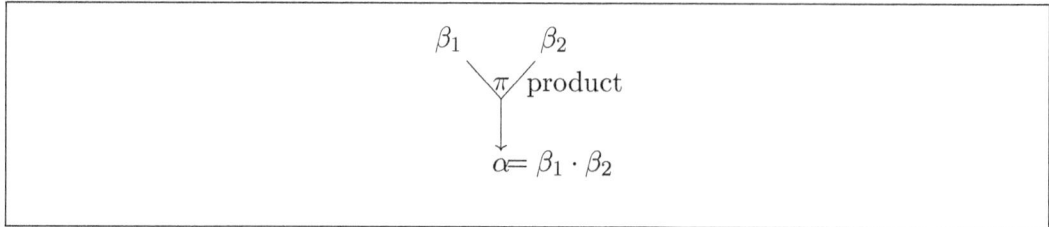

Figure 11

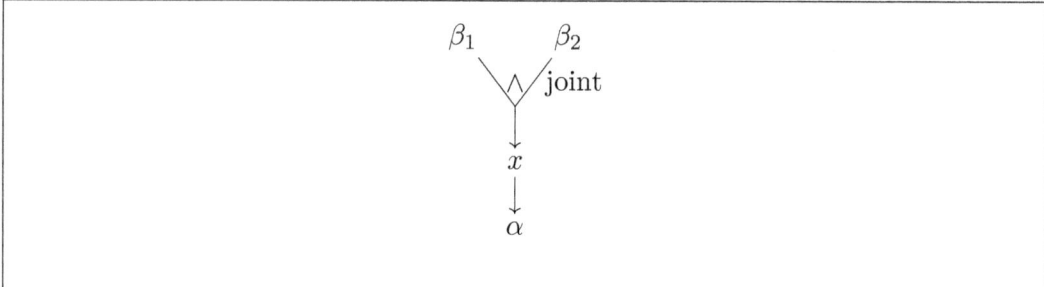

Figure 12

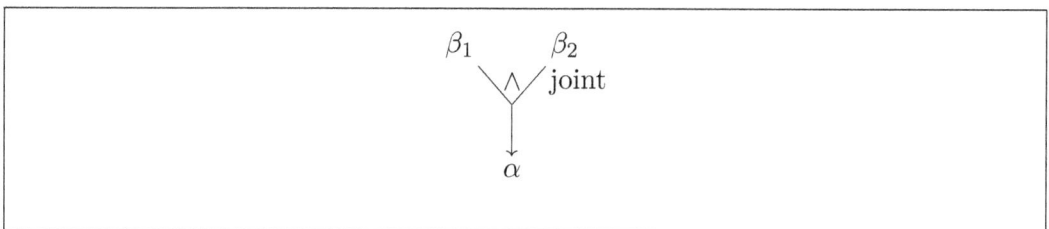

Figure 13

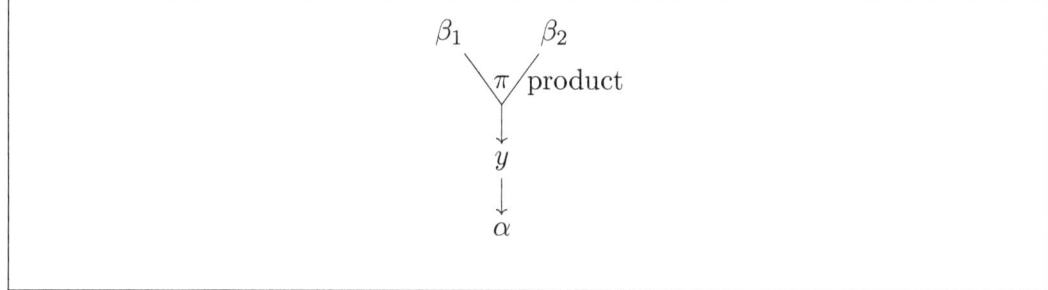

Figure 14

From (3), one can see that only if all $f(\beta_j) = 1$, do we get $f(\alpha) = 0$.

So we are using the equation below for the joint attacks of Fig. 10.

$$\alpha = \min(1, \sum_j 1 - \beta_j)$$

We make some comments about this equation.

(a) First, note that this equational truth-table is definable in Łukasiewicz logic [23]. In this logic, propositions get value in $[0, 1]$, 1 represents truth and 0 represents falsity. The truth-tables for $\neg$ and $\to$ are:

$$\neg X = 1 - X$$
$$X \to Y = \min(1, 1 - X + Y)$$

Therefore,

$$X \to \neg Y = \min(1, 1 - X + 1 - Y)$$

Let φ be a new connective operating on a non-empty list $[X_1, \ldots, X_n]^7$ defined as follows.

$$\varphi([X_1]) = \neg X_1$$
$$\varphi([X_1, X_2]) = X_2 \to \varphi([X_1])$$

By induction, assume $\varphi(X_1, \ldots, X_n)$ is definable and satisfies

$$\varphi(X_1, \ldots, X_n) = min(1, \sum_i (1 - X_i))$$

Then

$$X_{n+1} \to \varphi(X_1, \ldots, X_n) = \min(1, 1 - X_{n+1} + \varphi(X_1, \ldots, X_n))$$
$$= \min(1, \sum_{i=1}^{n+1} (1 - X_i))$$
$$= \varphi(X_1, \ldots, X_{n+1})$$

Note that $\varphi(X_1) = \min(1, 1 - X_1) = \neg X_1$. So we have

$$\varphi(X_1) = \neg X_1$$
$$\varphi(X_1, \ldots, X_n, X_{n+1}) = X_{n+1} \to \varphi(X_1, \ldots, X_n)$$

(b) Also note that argumentation networks with a formula of the type of φ just defined for joint attacks are not definable through the traditional Dung semantics. This gives new semantics. Consider Fig. 15.

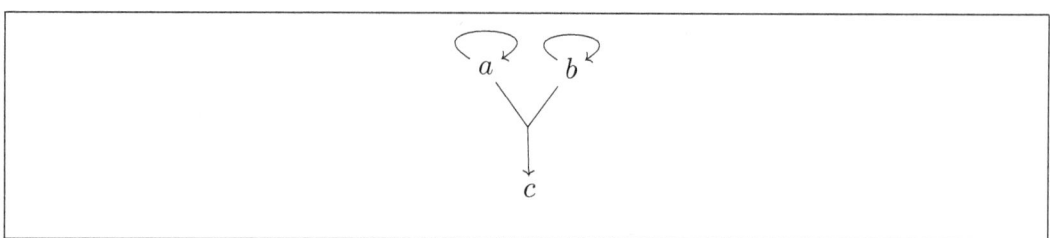

Figure 15

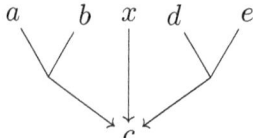

$$c = \min(1, 1 - a + 1 - b) \times \min(1, 1 - x) \times \min(1, 1 - d + 1 - e)$$
$$= \min(1, 1 - a + 1 - b) \times (1 - x) \times \min(1, 1 - d + 1 - e)$$

Figure 16

Use only φ. We get $a = \frac{1}{2}$, $b = \frac{1}{2}$, $c = \min(1, \frac{1}{2} + \frac{1}{2}) = 1$. So $c =$ "in".

The rationale behind this semantics is as follows.

We reject c if there are joint attacks on c where all the attackers are "in". If some attackers are "undecided", then so is c. However, if too many attackers are "undecided" (and remember this is a consortium joint attack where too many members of the consortium are undecided), then we disregard the attack and let $c =$ "in".

To get a better idea of how Eq_{inv} with Łukasiewicz joint attacks works, compare Figures 16 and 17.

The equation in Fig. 17 is given by Eq_{inv}, for β_1 being the joint attack of $\{a, b\}$, β_2 being the attack of x, and β_3 being the joint attack of $\{d, e\}$. The joint attacks of β_1 and β_3 are calculated each according to Equation 3 (see page 15). So going back to Fig. 16, under the above considerations we get

$$c = \min(1, 1 - a + 1 - b) \times (1 - x) \times \min(1, 1 - d + 1 - e).$$

---

[7]For example conjunction is an operator that can be seen to be operating on a list, where $\bigwedge([X]) = X$ and $\bigwedge([X_1, \ldots, X_n]) = X_n \bigwedge([X_1, \ldots, X_{n-1}])$.

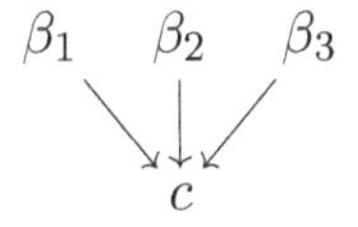

Bayesian Argumentation Networks

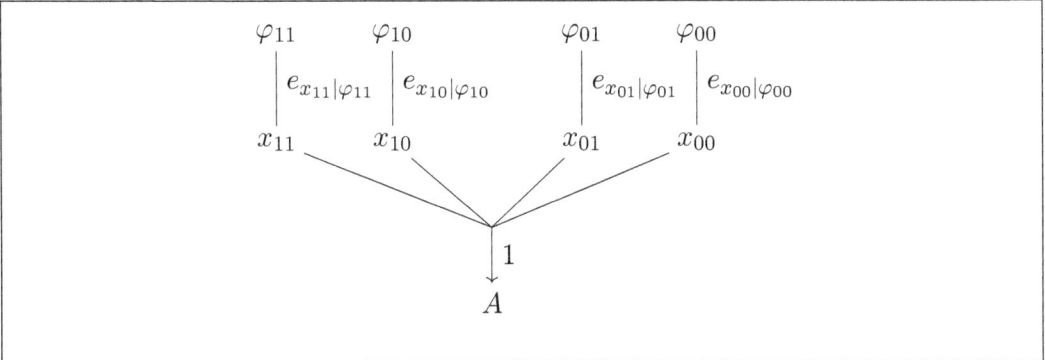

$$c = \min(1, 1-\beta_1) \times \min(1, 1-\beta_2) \times \min(1, 1-\beta_3)$$
$$= (1-\beta_1) \times (1-\beta_2) \times (1-\beta_3)$$
$$= \prod_i (1-\beta_i)$$

Figure 17

Figure 18: A complex configuration of joint attacks with transmission factors.

2.2 Combining It All

Consider Fig. 18.

The equational approach will give a solution f to this configuration as

$$f(x_{ij}) = 1 - e_{ij} \times f(\varphi_{ij})$$

and

$$f(A) = \min\left(1, \sum(1 - f(x_{ij}))\right)$$
$$f(A) = \min\left(1, \sum(1 - (1 - e_{ij} \times f(\varphi_{ij})))\right)$$
$$f(A) = \min\left(1, \sum e_{ij} \times f(\varphi_{ij})\right)$$

Now look again at Fig. 3. If we can instantiate φ_{ij} by an appropriate attack

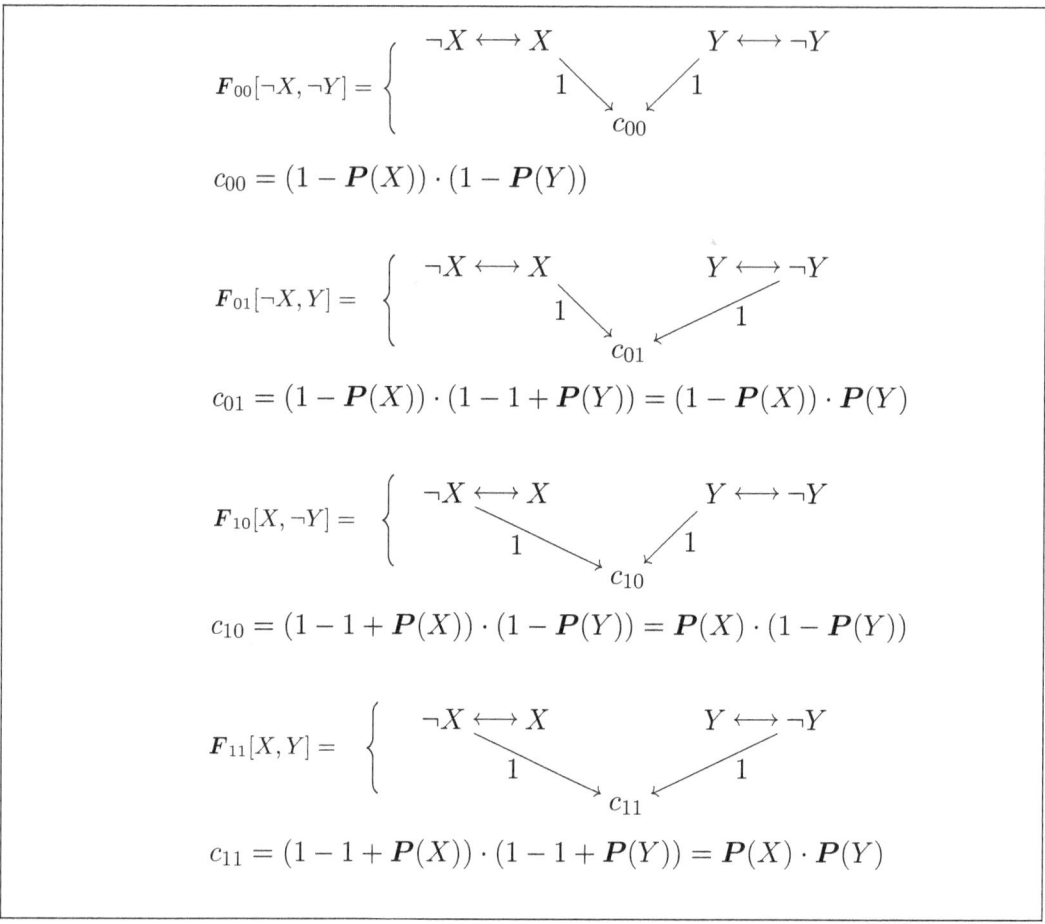

Figure 19

formation F_{ij} such that

$$f(\varphi_{ij}) = f(F_{ij}) = P^{\pm}(X) \times P^{\pm}(Y)$$

then we have implemented that figure as Fig. 18.

This is easy to do. Look at Fig. 19 and remember that since X, $\neg X$, Y and $\neg Y$ are end points attacking each other respectively, we can give them arbitrary probabilities! In Fig. 19, c_{ij} is the output point (realising $f(\varphi_{ij})$), and X and Y are given probabilities $P(X)$ and $P(Y)$, respectively. This gives probabilities $P(\neg X) = 1 - P(X)$ and $P(\neg Y) = 1 - P(Y)$.

Remark 2. *Note that the above implementation required two types of attacks. The product attack of Fig. 11 and the new joint attack of Fig. 10, namely*

$$\alpha = \min(1, \textstyle\sum_j (1 - \beta_j)).$$

We know that the new joint attack can be expressed in Łukasiewicz infinite-valued logic. Therefore in the extension of this logic with product conjunctions, namely with the additional connective $*$:[8]

$$x * y = x \cdot y$$

we can implement Bayesian networks!

3 Formal Definitions

This section will describe the formal machinery of Bayesian Argumentation Networks (BANs) and the mechanism to translate a Bayesian network into a BAN.

Definition 1 (Bayesian Argumentation Network (BAN)).

a. *A BAN has the form $\mathbb{B} = \langle S, R, \mathbf{e} \rangle$, where S is a non-empty set of arguments, $R \subseteq (2^S - \varnothing) \times S$ is the attack relation between non-empty subsets of S and an element of S. For each pair $(H, x) \in R$, such that $H \subseteq S$ and $x \in S$, $\mathbf{e}$ is a transmission function giving each h in the pair (H, x) a real value $\mathbf{e}(H, x, h) \in [0, 1]$.*

We can describe this situation in Fig. 20, where $(H, x) \in R$, $H = \{h_1, \ldots, h_k\}$ and $e_i = \mathbf{e}(H, x, h_i)$.

The general attack configuration of a node is depicted in Fig. 21, where $H^1 = \{h_1^1, \ldots, h_{k_1}^1\}, \ldots, H^i, \ldots, H^m = \{h_1^m, \ldots, h_{k_m}^m\}$ are all attackers of the node x. In such configuration $e_{ij} = \mathbf{e}(H^i, x, h_{ij})$.

b. *Let f be a function from S into $[0, 1]$. The equation associated with f and the configuration of Fig. 21 is*

$$f(x) = \prod_{j=1}^{m} \min(1, \sum_{i=1}^{k_j} (1 - e_{ij} \cdot f(h_i^j))) \qquad (4)$$

c. *A solution to equation (4) of item b. is called an extension to $\mathbb{B}$.*

---

[8]See [10] for details.

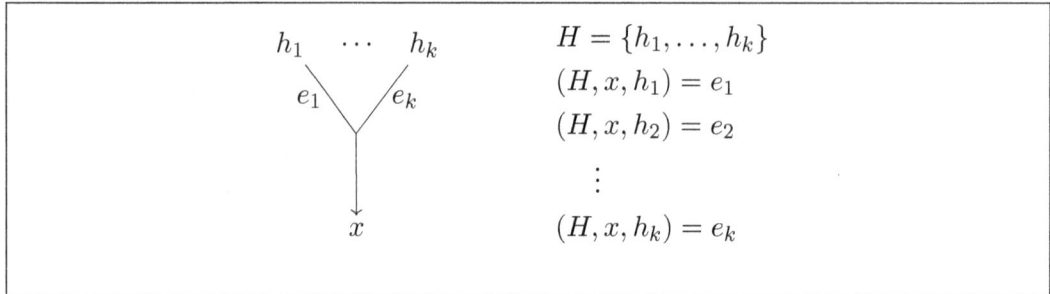

Figure 20

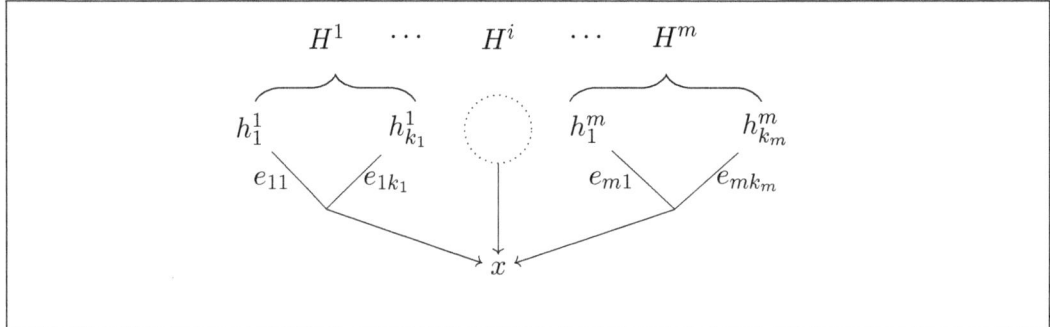

Figure 21

Example 1. *Consider Fig. 22. In this figure we have:*

$$\begin{aligned}
S &= \{a,b,c,d,e,x\} \\
R &= \{(\{a,b\},c),(\{x\},c),(\{d,e\},c)\} \\
e(\{a,b\},c,a) &= e_1 \\
e(\{a,b\},c,b) &= e_2 \\
e(\{x\},c,x) &= e_3 \\
e(\{d,e\},c,d) &= e_4 \\
e(\{d,e\},c,e) &= e_5
\end{aligned}$$

The equation for c is

$$\begin{aligned}
f(c) &= \min(1, 1 - e_1 \cdot f(a) + 1 - e_2 \cdot f(b)) \times \\
&\quad \min(1 - e_3 \cdot f(x)) \times \\
&\quad \min(1, 1 - e_4 \cdot f(d) + 1 - e_5 \cdot f(e))
\end{aligned}$$

Take $e_i = 1$ and compare this with Fig. 16.

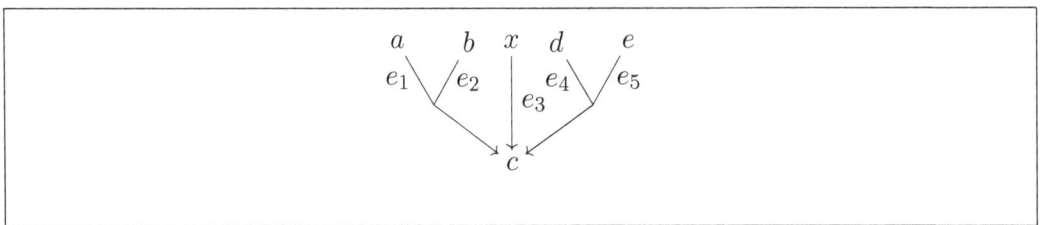

Figure 22

Definition 2. *A Bayesian network $\mathcal{N}$ has the form $\langle S, \varrho, \boldsymbol{P} \rangle$ where S is a set of nodes, $\varrho \subset S \times S$ is an **acyclic** dependence relation, and $\boldsymbol{P}$ gives probability distributions based on (S, ϱ) defined below. The symbol Ψ_x will denote the set of parents of the node x in $\mathcal{N}$, i.e., $\Psi_x = \{y \mid (y, x) \in \varrho\}$.*

- *The elements of S are considered Boolean variables which can be in only one of two states either true $= \top = 1$, or false $= \bot = 0$.*

- *If $x \in S$ is a source node, i.e., $\Psi_x = \varnothing$, then $\boldsymbol{P}(x) \in [0,1]$ denotes the probability that $x = \top$. $\boldsymbol{P}(\neg x) = 1 - \boldsymbol{P}(x)$ denotes the probability that $X = \bot$.*

- *If $\Psi_x \neq \varnothing$, then $\boldsymbol{P}$ gives the conditional probability of x on $\Psi_x = \{y_1, \ldots, y_k\}$, denoted by $\boldsymbol{P}(x \mid \Psi_x)$. This means the following:*

    a. *First for each $q \in S$, consider a new atom letter denoted by $\neg q$. Read $q^0 \stackrel{def}{=} \neg q$ and $q^1 \stackrel{def}{=} q$.*

    b. *For each $\varepsilon \in 2^k$ (a vector of numbers $\varepsilon(i)$, $1 \leq i \leq k$ from $\{0,1\}$), we look at the option*
    $$\overrightarrow{y}(\varepsilon) = \wedge_i y_i^{\varepsilon(i)}$$
    where we read y_i^0 as $\neg y_i$ or $y_i = \bot$ and y_i^1 as y_i or $y_i = \top$.

    c. *We can now state what $\boldsymbol{P}(x \mid \Psi_x)$ is. $\boldsymbol{P}(x \mid \Psi_x)$ gives values $\boldsymbol{P}(x, \Psi_x, \varepsilon) \in [0, 1]$, for each $\varepsilon \in 2^k$.*

- *Let x be a node such that $\Psi_x \neq \varnothing$. Assuming that the probabilities $\boldsymbol{P}(y_i)$, for $y_i \in \Psi_x$, are known, the probability $\boldsymbol{P}(x)$ of x can be calculated as*
$$\boldsymbol{P}(x) = \sum_\varepsilon \boldsymbol{P}(x, \Psi_x, \varepsilon) \times \boldsymbol{P}(\overrightarrow{y}(\varepsilon))$$
where $\boldsymbol{P}(\overrightarrow{y}(\varepsilon))$ is $\prod_i \boldsymbol{P}(y_i)^{\varepsilon(i)}$ and $\boldsymbol{P}(y_i)^0 = 1 - \boldsymbol{P}(y_i)$ and $\boldsymbol{P}(y_i)^1 = \boldsymbol{P}(y_i)$. Once we know $\boldsymbol{P}(x)$, we also know $\boldsymbol{P}(\neg x) = 1 - \boldsymbol{P}(x)$.

Remark 3. *Note that certainly $\boldsymbol{P}(x) \geq 0$, but also that $\boldsymbol{P}(x) \leq \sum_\varepsilon \boldsymbol{P}(\overrightarrow{y}(\varepsilon)) = 1$, since all $\boldsymbol{P}(y_i)$ are probabilities.*

Remark 4. *Note that Definition 2 is restricted to Boolean values. Variables can be either in state $1 = $ true or in state $0 = $ false. So any conditional probability for a variable x depending on a variable y needs to give real numbers for each state of y, see condition c. of Definition 2. This is similar to Pearl's definition in [19]. Note however that we propagate values in the direction of the arrows, i.e., towards descendant nodes. Pearl allows for updating of probabilities in both directions (ancestors and descendant nodes) at any point in the network (see Section 2.2.3 of [19]). As we are dealing with argumentation networks, the restriction to boolean variables is more natural and we need not be concerned about it. However, the propagation of updates in both directions is important and should be investigated not only in order to be more faithful in translating Bayesian networks into argumentation networks but even without any connection with Bayesian networks. Just by looking at traditional Dung networks we may wish to insist on a certain status for an argument (i.e., "in", "out" or "undecided") and propagate this result in both directions. Our work on Bayesian networks may give us ideas on how to do that, not only in the case of Bayesian Argumentation Networks (i.e., reflecting their behaviour) but perhaps by also taking advantage of the argumentation environment in developing an update theory applicable in argumentation in general, and not just restricted to the context of translated Bayesian networks. This requires extensive research and we leave it as future work.*

Example 2. *In order to illustrate Definition 2, we consider the network in Fig. 1 once more, with the probability values given in Section 2. We have $\boldsymbol{P}(X) = 0.01$, so $\boldsymbol{P}(\neg X) = 0.99$. $\boldsymbol{P}(Y) = 0.02$, so $\boldsymbol{P}(\neg Y) = 0.98$.*

$\Psi_A = \{X, Y\}$. We have that $\boldsymbol{P}(A, \Psi_A, X \wedge Y) = 0.95$; $\boldsymbol{P}(A, \Psi_A, X \wedge \neg Y) = 0.94$; $\boldsymbol{P}(A, \Psi_A, \neg X \wedge Y) = 0.29$; and $\boldsymbol{P}(A, \Psi_A, \neg X \wedge \neg Y) = 0.001$. According to

Definition 2:

$$\begin{aligned} \boldsymbol{P}(A) &= \boldsymbol{P}(A, \Psi_A, X \wedge Y) \times \boldsymbol{P}(X) \times \boldsymbol{P}(Y) + \\ &\quad \boldsymbol{P}(A, \Psi_A, X \wedge \neg Y) \times \boldsymbol{P}(X) \times \boldsymbol{P}(\neg Y) + \\ &\quad \boldsymbol{P}(A, \Psi_A, \neg X \wedge Y) \times \boldsymbol{P}(\neg X) \times \boldsymbol{P}(Y) + \\ &\quad \boldsymbol{P}(A, \Psi_A, \neg X \wedge \neg Y) \times \boldsymbol{P}(\neg X) \times \boldsymbol{P}(\neg Y) \end{aligned}$$

$$\begin{aligned} \boldsymbol{P}(A) &= 0.95 \times 0.01 \times 0.02 + \\ &\quad 0.94 \times 0.01 \times 0.98 + \\ &\quad 0.29 \times 0.99 \times 0.02 + \\ &\quad 0.001 \times 0.99 \times 0.98 \end{aligned}$$

$$\boldsymbol{P}(A) \approx 0.016$$

We then get $\boldsymbol{P}(\neg A) = 1 - \boldsymbol{P}(A) = 0.984$. The values of $\boldsymbol{P}(U)$ (resp., $\boldsymbol{P}(\neg U)$) and $\boldsymbol{P}(W)$ (resp., $\boldsymbol{P}(\neg W)$) are calculated in a similar way.

Definition 3. We now translate any Bayesian network $\mathcal{N} = \langle S, \varrho, \boldsymbol{P} \rangle$ into a Bayesian Argumentation Network $\langle A, R, \boldsymbol{e} \rangle$.

a. Assume the elements of S to be positive atoms of the form $\{q_i\}$. Let $\bar{S}$ be a new set of atoms of the form $\bar{S} = \{\bar{q} \mid q \in S\}$ and let $A_0 = S \cup \bar{S}$.

b. For any $x \in S$ such that Ψ_x has k elements, define two sets of new atoms

$$\begin{aligned} C(x, \Psi_x) &= \{c(x, \Psi_x, \varepsilon) \mid \varepsilon \in 2^k\} \text{ and} \\ D(x, \Psi_x) &= \{d(x, \Psi_x, \varepsilon) \mid \varepsilon \in 2^k\} \end{aligned}$$

Let $A = A_0 \cup \bigcup_{x \in S} (C(x, \Psi_x) \cup D(x, \Psi_x))$.

c. We now define R on A.

   (a) Have $(\{\bar{q}\}, q)$ and $(\{q\}, \bar{q})$ be in R.
   (b) For any $x \in S$ such that Ψ_x has k elements, let $(\{c(x, \Psi_x, \varepsilon)\}, d(x, \Psi_x, \varepsilon))$ be in R and let $(\{y_i^{1-\varepsilon(i)}\}, c(x, \Psi_x, \varepsilon))$ be in R.
   (c) Let $(D(x, \Psi_x), x)$ be in R.

d. We now define $\boldsymbol{e}$. We let

$$\boldsymbol{e}(\{c(x, \Psi_x, \varepsilon)\}, d(x, \Psi_x, \varepsilon), c(x, \Psi_x, \varepsilon)) = \boldsymbol{P}(x, \Psi_x, \varepsilon)$$

Theorem 1. *Let $\mathcal{N} = \langle S, \varrho, \boldsymbol{P}\rangle$ be a Bayesian network and let $\langle A, R, \boldsymbol{e}\rangle$ be its (argumentation) translation. Let the source nodes of $\mathcal{N}$ be the set $\Omega \subseteq S$ and let f be a solution extension of $(A, R, \boldsymbol{e})$ such that $f(w) = \boldsymbol{P}(w)$, for $w \in \Omega$. Then for every $s \in S$, $f(s) = \boldsymbol{P}(s)$ (remember that $S \subseteq A$).*

Proof. The proof is done by induction on the distance (level) of nodes from Ω.

Level 0: nodes $w \in \Omega$.

Level $n+1$: nodes s such that all nodes in Ψ_s are of level up to n with at least one of them being of level n.

Every node in S has a unique level because (S, ϱ) is acyclic. So we prove by induction on the level of a node s that $f(s) = \boldsymbol{P}(s)$.

Consider a node s of level $n+1$, such as the one in Fig. 23. Its translation into $(A, R, \boldsymbol{e})$ is Fig. 24. The computation of $\boldsymbol{P}(s)$ in the Bayesian network is

$$\boldsymbol{P}(s) = \sum_{\varepsilon \in 2^k} \boldsymbol{P}(s, y_s, \varepsilon) \times \prod_i \boldsymbol{P}(y_i)^{\varepsilon(i)}$$

Our inductive assumption is that $\boldsymbol{P}(y_i) = f(y_i)$. We want to show that $\boldsymbol{P}(s) = f(s)$. Let us calculate $f(s)$ from Fig. 24. First we have

$$\begin{aligned} f(y_i) &= \boldsymbol{P}(y_i) \\ f(\neg y_i) &= f(y_i^0) = 1 - f(y_i) \end{aligned}$$

Thus,

$$\begin{aligned} f(c(s, y_s, \varepsilon)) &= \prod_i (1 - f(y_i)^{1-\varepsilon(i)}) \\ &= \prod_i f(y_i^{\varepsilon(i)}) = \prod_i \boldsymbol{P}(y_i)^{\varepsilon(i)} \end{aligned}$$

So

$$\begin{aligned} f(d(s, y_s, \varepsilon)) &= 1 - \boldsymbol{e}(s, y_s, c(s, y_s, \varepsilon)) \times f(c(s, y_s, \varepsilon)) \\ &= 1 - \boldsymbol{P}(s, y_s, \varepsilon) \times \prod_i \boldsymbol{P}(y_i)^{\varepsilon(i)} \end{aligned}$$

Therefore,

$$\begin{aligned} f(s) &= \min(1, \sum_\varepsilon (1 - f(d(s, y_s, \varepsilon))) \\ &= \min(1, \sum_\varepsilon \boldsymbol{P}(s, y_s, \varepsilon) \times \prod_i \boldsymbol{P}(y_i)^\varepsilon(i)) \\ &= \boldsymbol{P}(s) \end{aligned}$$

□

Remark 5. *Note that the translation of a Bayesian network uses only a restricted fragment of Definition 1, where each node s satisfies for all y_s^i and for all y_s^j:*

$(y_s^i, s) \in R$ *and* $(y_s^j, s) \in R$ *implies either y_s^i and y_s^j are singleton sets or $y_s^i = y_s^j$*

This means that translated Bayesian networks do not have the combination of joint attacks and single attacks such as the one in Fig. 22. Nodes either have a unique joint attack or zero or more attacks by individual nodes.

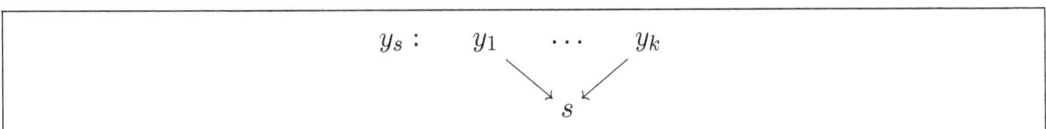

Figure 23: A node of level $n+1$.

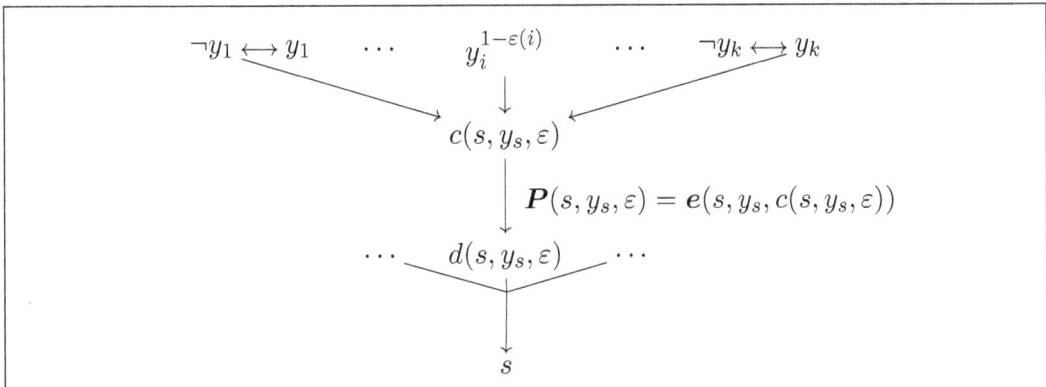

Figure 24: A node of level $n+1$.

Remark 6. a. Note that the class of Bayesian Argumentation Networks contains more networks than just translated images of original Bayesian networks. These are indeed a type of argumentation networks inspired by looking at Bayesian networks. Note also that images of Bayesian networks are faithful and can translated back into Bayesian networks.

b. Another important point to observe is that we are not imposing an argumentation structure on top of a Bayesian network, thus analysing the Bayesian network from an argumentation point of view. Such methods are common in the argumentation community. This will be discussed further in Section 6.

4 A Comprehensive Translation Example

In this section, we show in detail how a Bayesian network can be translated into a Bayesian Argumentation Network. For this, we will use the network of Fig. 1 as an example. The translation starts with the source nodes of the Bayesian network.

Consider a node α and its attackers $Att(\alpha) = \{\beta_1, \ldots, \beta_k\}$. Assume that the nodes in $Att(\alpha)$ are all source nodes, without any attackers. When β_i is a source node, it is given an initial probability $\boldsymbol{P}(\beta_i)$. We model this in a BAN by adding a new node $\neg \beta_i$ for each β_i such that β_i and $\neg \beta_i$ attack each other. Now, in the new network, β_i can assume any initial probability $\boldsymbol{P}(\beta_i)$ we want, with $\neg \beta_i$ obtaining $\boldsymbol{P}(\neg \beta_i) = 1 - \boldsymbol{P}(\beta_i)$. In order to represent all possible conjunctive expressions of the form

$$e_{j=0}^{2^k - 1} = \wedge_{i_j} \beta_{i_j}^{\pm}$$

where $\beta_{i_j}^+ = \beta_{i_j}$ and $\beta_{i_j}^- = \neg \beta_{i_j}$, we need to create 2^k intermediate points c_j, whose attackers are all $\beta_i, \neg \beta_i$, such that $e_j \models \beta_i$, and $e_j \models \neg \beta_i$, respectively. Note that the value of each c_j will now correspond to a particular product

$$\boldsymbol{P}(c_j) = \prod_{i_j} p_{i_j}$$

where

$$p_{i_j} = \begin{cases} 1 - \boldsymbol{P}(\beta_i), & \text{if } e_j \models \beta_i \\ \boldsymbol{P}(\beta_i), & \text{if } e_j \models \neg \beta_j \end{cases}$$

Because we also want to represent transmission values, we now need to duplicate each intermediate point c_j with a corresponding x_j and have c_j attack x_j with transmission factor e_j.

In order to illustrate this, let us recall the Bayesian network of Fig. 1.

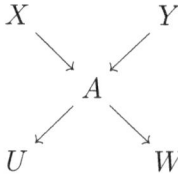

Consider the node A, with attackers X and Y. We first add points $\neg X$ and $\neg Y$ such that each of X and $\neg X$ and Y and $\neg Y$ attack each other. If we give value $\boldsymbol{P}(X)$ to X and value $\boldsymbol{P}(Y)$ to Y, then $\neg X$ will get value $1 - \boldsymbol{P}(X)$ and $\neg Y$ will get value $1 - \boldsymbol{P}(Y)$.

We now add points c_{00}, c_{01}, c_{10} and c_{11} corresponding to the conjunctive expressions

$$\begin{aligned}
e_0 : c_{00} &= X \wedge Y \\
e_1 : c_{01} &= X \wedge \neg Y \\
e_2 : c_{10} &= \neg X \wedge Y \\
e_3 : c_{11} &= \neg X \wedge \neg Y
\end{aligned}$$

Note that $e_0 \models X$ and $e_0 \models Y$, so we add attacks from X and Y into c_{00} and do the same for e_1–e_3.

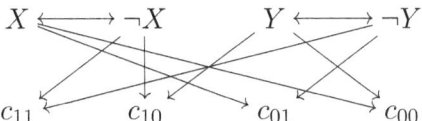

We can now see that the probabilities of c_{ij} are given as follows

$$\begin{aligned}
\boldsymbol{P}(c_{00}) &= (1 - \boldsymbol{P}(X)) \cdot (1 - \boldsymbol{P}(Y)) \\
\boldsymbol{P}(c_{01}) &= (1 - \boldsymbol{P}(X)) \cdot (1 - 1 + \boldsymbol{P}(Y)) = (1 - \boldsymbol{P}(X)) \cdot \boldsymbol{P}(Y) \\
\boldsymbol{P}(c_{10}) &= (1 - 1 + \boldsymbol{P}(X)) \cdot (1 - \boldsymbol{P}(Y)) = \boldsymbol{P}(X) \cdot (1 - \boldsymbol{P}(Y)) \\
\boldsymbol{P}(c_{11}) &= (1 - 1 + \boldsymbol{P}(X)) \cdot (1 - 1 + \boldsymbol{P}(Y)) = \boldsymbol{P}(X) \cdot \boldsymbol{P}(Y)
\end{aligned}$$

Remember the initial parameters given for this network:

$$\begin{aligned}
\boldsymbol{P}(X) &= 0.01 \\
\boldsymbol{P}(\neg X) &= 0.99 \\
\boldsymbol{P}(Y) &= 0.02 \\
\boldsymbol{P}(\neg Y) &= 0.98
\end{aligned}$$

This gives us

$$\begin{aligned}
\boldsymbol{P}(c_{00}) &= 0.99 \times 0.98 = 0.9702 \\
\boldsymbol{P}(c_{01}) &= 0.99 \times 0.02 = 0.0198 \\
\boldsymbol{P}(c_{10}) &= 0.01 \times 0.98 = 0.0098 \\
\boldsymbol{P}(c_{11}) &= 0.01 \times 0.02 = 0.0002
\end{aligned}$$

Now for each node c_{ij}, we add an additional node x_{ij}, so that we can incorporate the transmission factors e_{ij}.

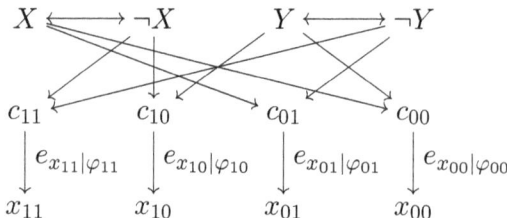

The resulting probabilities of the nodes x_{ij} are then given by the equations

$$P(x_{00}) = 1 - e_{x_{00}|\varphi_{00}} \cdot c_{00}$$
$$P(x_{01}) = 1 - e_{x_{01}|\varphi_{01}} \cdot c_{01}$$
$$P(x_{10}) = 1 - e_{x_{10}|\varphi_{10}} \cdot c_{10}$$
$$P(x_{11}) = 1 - e_{x_{11}|\varphi_{11}} \cdot c_{11}$$

With our initial parameters, we have that

$$e_{x_{00}|\varphi_{00}} = 0.001$$
$$e_{x_{01}|\varphi_{01}} = 0.29$$
$$e_{x_{10}|\varphi_{10}} = 0.94$$
$$e_{x_{11}|\varphi_{11}} = 0.95$$

and hence

$$P(x_{00}) = 1 - 0.001 \times 0.9702 = 0.9990$$
$$P(x_{01}) = 1 - 0.29 \times 0.0198 = 0.9942$$
$$P(x_{10}) = 1 - 0.94 \times 0.0098 = 0.9907$$
$$P(x_{11}) = 1 - 0.95 \times 0.0002 = 0.9998$$

The x_{ij} jointly attack A:

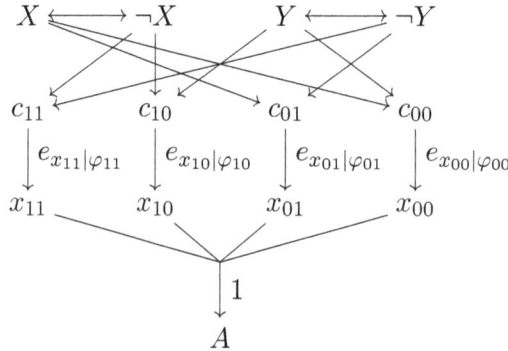

This finally gives us the value of the probability of node A as

$$\begin{aligned} P(A) &= \min\{1, \sum_{ij}(1-P(x_{ij}))\} \\ &= \min\{1, 0.001 + 0.0058 + 0.0093 + 0.0002\} \\ &\approx 0.016 \end{aligned}$$

as before. Therefore, $P(\neg A) \approx 0.984$.

Now to proceed to the next level all we need to do is to add a complementary node to A, $\neg A$, such that A and $\neg A$ attack each other. This will give us $P(\neg A) = 1 - P(A)$. As before, we also need new intermediate nodes w_A, $w_{\neg A}$, u_A and $u_{\neg A}$ to incorporate transmission factors.

The original attack of A on W is then realised by the joint attack of the new intermediate nodes $w_{\neg A}$ and w_A and the attack of $\neg A$ on U is realised by the joint attack of the new intermediate nodes $u_{\neg A}$ and u_A.

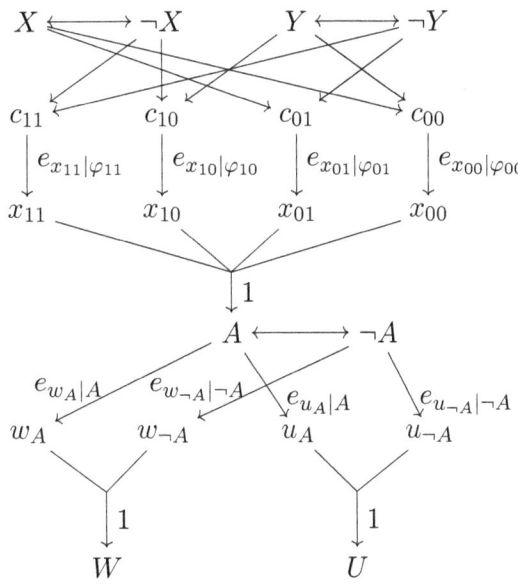

The values of w_A, $w_{\neg A}$, u_A and $u_{\neg A}$ are calculated as follows.

$$\begin{aligned} P(w_A) &= 1 - e_{w_A|A} \cdot P(A) \\ P(w_{\neg A}) &= 1 - e_{w_{\neg A}|\neg A} \cdot P(\neg A) \\ P(u_A) &= 1 - e_{u_A|A} \cdot P(A) \\ P(u_{\neg A}) &= 1 - e_{u_{\neg A}|\neg A} \cdot P(\neg A) \end{aligned}$$

Since

$$e_{w_A|A} = 0.7$$
$$e_{w_{\neg A}|\neg A} = 0.01$$
$$e_{u_A|A} = 0.9$$
$$e_{u_{\neg A}|\neg A} = 0.05$$

We get

$$\boldsymbol{P}(w_A) = 1 - 0.7 \times 0.016 \approx 0.988$$
$$\boldsymbol{P}(w_{\neg A}) = 1 - 0.01 \times 0.984 \approx 0.990$$
$$\boldsymbol{P}(u_A) = 1 - 0.9 \times 0.016 \approx 0.985$$
$$\boldsymbol{P}(u_{\neg A}) = 1 - 0.05 \times 0.984 \approx 0.950$$

We can finally calculate the values of W and U.

$$\boldsymbol{P}(W) = \min\{1, 1 - \boldsymbol{P}(w_{\neg A}) + 1 - \boldsymbol{P}(w_A)\} = \min\{1, 0.009 + 0.0112\} \approx 0.021$$
$$\boldsymbol{P}(U) = \min\{1, 1 - \boldsymbol{P}(u_{\neg A}) + 1 - \boldsymbol{P}(u_A)\} = \min\{1, 0.049 + 0.0144\} \approx 0.063$$

These are exactly the same values we had before.

5 Complexity Discussion

We need to consider two aspects involved in the complexity arising from the translation of a Bayesian network into a Bayesian Argumentation Network. The first one is related to the translation itself whereas the second is related to the actual computation of the node values.

It is easy to see that the translation of a Bayesian network into a Bayesian Argumentation Network according to Definition 3 results in the creation of many new nodes. This addition of nodes is linear on the number of nodes of the original Bayesian network (item a. of Definition 3) and exponential on the number of ancestors of each node of the original Bayesian network (item b. of Definition 3).

The effect of the increase in the number of nodes in the actual computation of node values is deceptive.

Although the translation does incur in the addition of many new nodes, the complexity of the actual calculation of the node values remains basically the same.

The nodes added in a Bayesian Argumentation Network simply encode explicitly the intermediate calculations that would otherwise be implicitly done in the original Bayesian network.

In conclusion, a Bayesian Argumentation Network simply explicitly encodes as nodes the calculations that would have to be done anyway in the original Bayesian network. The new nodes name key values when using the conditional probability tables in the original calculations of the Bayesian network. Therefore there is no significant additional cost.

6 Related Work

Let us compare with several related papers. We start with Vreeswijk's "Argumentation in bayesian belief networks" [27]. An important point to observe is that we are not using a meta-level device of extracting/identifying arguments and a notion of attack on the arguments from the Bayesian network and thus obtaining an argumentation network as is done in [27].

Vreeswijk's approach does not give us a direct connection/translation between Bayesian networks and argumentation networks. It is more akin to imposing an argumentation structure on top of a Bayesian network, thus analysing the Bayesian network from an argumentation point of view. Indeed Vreeswijk sees his approach [27] as

> "a proposal to look at Bayesian belief networks from the perspective of argumentation. More specifically, I propose an algorithm that enables users to start an argumentation process within the context of an existing Bayesian belief network. ...with some imagination, the CPTs[9] of the above Bayesian network can be translated into the rule-base and evidence ...A next step towards argumentation is to chain rules into arguments. ...What remains to be done to obtain a full-fledged argument system, is to define an attack relation between pairs of arguments. To this end, I choose to define the notion of attack on the basis of two notions that are more elementary and (therefore) fall beyond the scope of a Dung-type argument system, viz. the notion of counterargument and the notion of strength of an argument. First I will discuss counter-arguments, and then I will discuss argument strength. ...
>
> **Definition 4 (Attack).** We say that argument a is attacked by argument b, written $a \leftarrow b$, if it satisfies the following two conditions:

---
[9] Conditional Probability Tables

1. Argument b is a counterargument of a sub-argument a' of a.

2. Argument b is stronger than argument a'."

Such methods are common in the argumentation community. For example, taking a logic program, forming arguments using this logic program and thus generating an argumentation network, and then proving equivalence of the logic program with the argumentation network (see for example [7, 28]).

The next related work we consider is the translation of various kinds of argumentation networks into Bayesian networks, as is done for example in [16]. The idea is beautifully described by the authors:

"This paper presents a technique with which instances of argument structures in the Carneades model can be given a probabilistic semantics by translating them into Bayesian networks. The propagation of argument applicability and statement acceptability can be expressed through conditional probability tables. This translation suggests a way to extend Carneades to improve its utility for decision support in the presence of uncertainty."

Note that [27] identifies some argumentation structure in Bayesian networks whilst [16] uses Bayesian networks to implement certain kinds of argumentation networks. There is similarity but the approaches are different. On the other hand, our approach is to faithfully represent Bayesian networks inside a newly motivated numerical argumentation network.

Finally, the approach used in [25, 24] is similar in spirit to the one used in [27]. The idea is to provide explanations of Bayesian networks in legal or medical domains using support graphs. The nodes of the support graph can be labelled to provide an argumentative interpretation of the Bayesian network.

7 Conclusions and Future Work

The perceptive reader from the Bayesian community may take no interest in the translation of Bayesian networks into argumentation. They will point out justifiably, that we are just translating the Bayesian algorithms into argumentation and then using these same algorithms under the guise of argumentation. This may be of interest from the abstract mathematical expressive power point of view, but that is all.

We would like to show that there are indeed other benefits to this translation both to the argumentation community and to the Bayesian community.

a. By interpreting Bayesian networks in some extension of Dung's networks and vice-versa, the argumentation community, who also deals with updates and change, stands to benefit from the updating algorithms in Bayesian networks. We can add nodes and change values of nodes and use Bayesian propagation to propagate the changes.

This needs to be studied further.

b. Bayesian networks are acyclic, they have difficulties with loops (see [26]). The equational approach in argumentation can deal with loops without any problems (see for instance the forthcoming special issue in the Journal of Logic and Computation on the Handling of Loops in Argumentation Networks and see [11]). We notice that Bayesian networks can adopt the equational approach [15] in case of loops and simply solve the equations which arise on the probabilities. A solution always exists by Brouwer's fixed-point theorem. We are in fact surprised that this approach has never been taken (to the best of our knowledge) by the Bayesian community. One can then use loop handling techniques from argumentation to propagate updates and changes in the (cyclic) Bayesian network by simply solving equations. By implementing Bayesian networks in argumentation under the equational approach we can allow for loops in Bayesian networks and still hopefully see a way to obtain results, again, this requires further research.

c. A third benefit to Bayesian networks is the possibility to develop proof theory for Bayesian networks. The extension of argumentation networks which hosts the translation of Bayesian networks is implemented in Łukasiewicz infinite-valued logic with product.[10] This Łukasiewicz logic has a proof theory (see [10]). We can therefore hope for the same for Bayesian networks. Again this needs to be studied.

In addition, the following needs to be investigated in detail:

a. Take an example of a cyclic Bayesian network, recognised as interesting in the Bayesian community, and translate it into argumentation. See how argumentation handles the loops and try to find suitable algorithms for handling cycles in Bayesian networks. These algorithms should now be independent of argumentation.

b. Develop algorithms of updating argumentation networks by looking at the examples of updating used in the Bayesian domain and the way they implement

---

[10]This is actually shown in the current paper (see Remark 2).

the updates. There are important papers investigating updates and revision of argumentation networks by central figures in the argumentation community, including [4, 22, 21, 5, 8, 3, 9]. Addressing these papers will be done in future work.

c. Identify proof theoretic queries in Bayesian networks and import proof theory from Łukasiewicz logic to model them.

References

[1] H. Barringer, D. M. Gabbay, and J. Woods. Temporal dynamics of support and attack networks. In D. Hutter and W. Stephan, editors, *Mechanizing Mathematical Reasoning*, 2005. LNCS, vol. 2605.

[2] Howard Barringer, Dov M. Gabbay, and John Woods. Temporal, numerical and meta-level dynamics in argumentation networks. *Argument & Computation*, 3(2-3):143–202, 2012.

[3] R. Baumann and G. Brewka. AGM meets abstract argumentation: Expansion and revision for dung frameworks. In *Proceedings of the Twenty-Fourth International Joint Conference on Artificial Intelligence, IJCAI 2015, Buenos Aires, Argentina, July 25-31, 2015*, pages 2734–2740, 2015.

[4] G. Brewka, S. Ellmauthaler, H. Strass, J. P. Wallner, and S. Woltran. Abstract dialectical frameworks revisited. In *Proceedings of the Twenty-Third International Joint Conference on Artificial Intelligence*, IJCAI '13, pages 803–809. AAAI Press, 2013.

[5] G. Brewka and S. Woltran. GRAPPA: A semantical framework for graph-based argument processing. In *ECAI 2014 - 21st European Conference on Artificial Intelligence, 18-22 August 2014, Prague, Czech Republic - Including Prestigious Applications of Intelligent Systems (PAIS 2014)*, pages 153–158, 2014.

[6] Gerhard Brewka and Stefan Woltran. Abstract dialectical frameworks. In *Principles of Knowledge Representation and Reasoning: Proceedings of the Twelfth International Conference, KR 2010*. AAAI Press, 2010.

[7] Martin Caminada, Samy Sá, Jo ao Alcântara, and Wolfgang DvoÅŹák. On the equivalence between logic programming semantics and argumentation semantics. *International Journal of Approximate Reasoning*, 58:87 – 111, 2015. Special Issue of the Twelfth European Conference on Symbolic and Quantitative Approaches to Reasoning with Uncertainty (ECSQARU 2013).

[8] S. Coste-Marquis, S. Konieczny, J-G. Mailly, and P. Marquis. On the revision of argumentation systems: Minimal change of arguments statuses. In *14th International Conference on Principles of Knowledge Representation and Reasoning (KR'14)*, Vienna, July 2014.

[9] M. Diller, A. Haret, T. Linsbichler, S. Rümmele, and S. Woltran. An extension-based approach to belief revision in abstract argumentation. In *Proceedings of the 24th Inter-*

national Conference on Artificial Intelligence, IJCAI'15, pages 2926–2932. AAAI Press, 2015.

[10] Francesc Esteva, Lluís Godo, and Enrico Marchioni. *Fuzzy Logics with Enriched Language*, chapter VIII, pages 627 – 711. Number 38 in Vol. 2. College Publications, London, 2011.

[11] D. M. Gabbay. The handling of loops in argumentation networks. *Journal of Logic and Computation*, 2014.

[12] D. M. Gabbay. Theory of semi-instantiation in abstract argumentation. *Logica Universalis*, To appear.

[13] D. M. Gabbay and O. Rodrigues. Probabilistic argumentation: An equational approach. *Logica Universalis*, pages 1–38, 2015.

[14] Dov M. Gabbay. Fibring argumentation frames. *Studia Logica*, 93(2-3):231–295, 2009.

[15] Dov M. Gabbay. Equational approach to argumentation networks. *Argument & Computation*, 3(2-3):87–142, 2012.

[16] M. Grabmair, T. F. Gordon, and D. Walton. Probabilistic semantics for the carneades argument model using bayesian networks. In *Proceedings of the 2010 Conference on Computational Models of Argument: Proceedings of COMMA 2010*, pages 255–266, Amsterdam, The Netherlands, The Netherlands, 2010. IOS Press.

[17] Kevin B. Korb and Ann E. Nicholson. *Bayesian Artificial Intelligence, Second Edition*. CRC Press, Inc., Boca Raton, FL, USA, 2nd edition, 2010.

[18] Søren Holbech Nielsen and Simon Parsons. A generalization of dung's abstract framework for argumentation: Arguing with sets of attacking arguments. In Nicolas Maudet, Simon Parsons, and Iyad Rahwan, editors, *ArgMAS*, volume 4766 of *Lecture Notes in Computer Science*, pages 54–73. Springer, 2006.

[19] J. Pearl. Fusion, propagation, and structuring in belief networks. *Artificial Intelligence*, 29(3):241–288, September 1986.

[20] J. Pearl. *Probabilistic Reasoning in Intelligent Systems: Networks of Plausible Inference*. Morgan Kaufmann Publishers Inc., San Francisco, CA, USA, 1988.

[21] S. Polberg and D. Doder. Probabilistic abstract dialectical frameworks. In Eduardo Fermé and João Leite, editors, *Logics in Artificial Intelligence: 14th European Conference, JELIA 2014, Funchal, Madeira, Portugal, September 24-26, 2014. Proceedings*, pages 591–599, Cham, 2014. Springer International Publishing.

[22] J. Pührer. Realizability of three-valued semantics for abstract dialectical frameworks. In *Proceedings of the 24th International Conference on Artificial Intelligence*, IJCAI'15, pages 3171–3177. AAAI Press, 2015.

[23] A. Rose. Formalisations of further $\aleph_0$-valued Łukasiewicz propositional calculi. *Journal of Symbolic Logic*, 43:207–210, 6 1978.

[24] S. T. Timmer, J-J Ch. Meyer, H. Prakken, S. Renooij, and B. Verheij. Explaining bayesian networks using argumentation. In *Symbolic and Quantitative Approaches to Reasoning with Uncertainty - 13th European Conference, ECSQARU 2015, Compiègne, France, July 15-17, 2015. Proceedings*, pages 83–92, 2015.

[25] S. T. Timmer, J-J Ch. Meyer, H. Prakken, S. Renooij, and B. Verheij. A structure-guided approach to capturing bayesian reasoning about legal evidence in argumentation. In *Proceedings of the 15th International Conference on Artificial Intelligence and Law*, ICAIL '15, pages 109–118, New York, NY, USA, 2015. ACM.

[26] A. L. Tulupyev and S.I. Nikolenko. Directed cycles in bayesian belief networks: Probabilistic semantics and consistency checking complexity. In A. Gelbukh, A. de Albornoz, and H. Terashima, editors, *MICAI 2005, Lectune notes in Artificial Intelligence 3789*, pages 214–223. Springer-Verlag Berlin Heidelberg, 2005.

[27] G. A. W. Vreeswijk. Argumentation in bayesian belief networks. In *Proceedings of the First International Conference on Argumentation in Multi-Agent Systems*, ArgMAS'04, pages 111–129, Berlin, Heidelberg, 2005. Springer-Verlag.

[28] Yining Wu, Martin Caminada, and Dov M. Gabbay. Complete extensions in argumentation coincide with 3-valued stable models in logic programming. *Studia Logica*, 93(2):383–403, 2009.

www.ingramcontent.com/pod-product-compliance
Lightning Source LLC
Chambersburg PA
CBHW081347040426
42450CB00015B/3338